T0321205

BASICS OF
STATISTICAL PHYSICS

Third Edition

BASICS OF
STATISTICAL PHYSICS

Third Edition

Harald J. W. Müller-Kirsten

University of Kaiserslautern, Germany

World Scientific

NEW JERSEY · LONDON · SINGAPORE · BEIJING · SHANGHAI · HONG KONG · TAIPEI · CHENNAI · TOKYO

Published by

World Scientific Publishing Co. Pte. Ltd.

5 Toh Tuck Link, Singapore 596224

USA office: 27 Warren Street, Suite 401-402, Hackensack, NJ 07601

UK office: 57 Shelton Street, Covent Garden, London WC2H 9HE

Library of Congress Control Number: 2022008145

British Library Cataloguing-in-Publication Data
A catalogue record for this book is available from the British Library.

BASICS OF STATISTICAL PHYSICS
Third Edition

ISBN 978-981-125-609-7 (hardcover)
ISBN 978-981-125-610-3 (ebook for institutions)
ISBN 978-981-125-611-0 (ebook for individuals)

For any available supplementary material, please visit
https://www.worldscientific.com/worldscibooks/10.1142/12830#t=suppl

Printed in Singapore

Contents

Preface to Third Edition

In the course of revision of the text many comments have been added, as well as numerous examples and exercises motivated by more recent developments. A section on black body thermal radiation is new, and is supplemented at the end by an introduction to thermal radiation of black holes.

Harald J.W. Müller–Kirsten

Preface to Second Edition

Apart from minor changes resulting from a revision of the text, the new features are the addition at the end of a chapter on the Boltzmann equation with applications, and some more examples inserted here and there, as well as additional references.

Harald J.W. Müller–Kirsten

Preface to First Edition

Statistical physics — more precisely statistical mechanics and statistical thermodynamics — is almost universally the last of the four basic theory courses for a degree in physics, and is usually taught after theory courses in mechanics, electrodynamics and quantum mechanics, thermodynamics proper, *i.e.* classical thermodynamics, being usually combined with this or covered in some form in connection with basic or experimental physics. Since the quantized form of statistical physics evolved naturally after the development of quantum mechanics in the first half of the twentieth century, this was a natural sequence of steps. Prior to this development, thermodynamics proper was the additional basic subject of courses and examinations. The heavy and detailed monograph of Mayer and Mayer [45], first published in 1940, shows how quickly the entire basis of quantum statistics was developed with numerous concurrent applications, so that soon appropriate texts appeared. One of these later texts was the book of Schrödinger [65]. This was the recommended text when the author had his first encounter with quantum statistics in 1956 in the second half of the third and final year for the *Bachelor of Science* degree (the Honours degree requiring an additional fourth year). Almost concurrently the extensive and relatively heavy monograph of Hill [33] appeared which became a leading text for many years. Thereafter, of course, more and more monographs were published, like the very readable second text of Hill [34] and the book of Rushbrooke [63], and today a large number of pedagogically arranged texts is available. Thus the author himself taught the subject repeatedly on the basis of the more extensive of the two widely known books of Reif [58], [59] together with some thermodynamics from Callen [10]. However later, motivated by the introduction of Bachelor degree courses in Germany, the author reconsulted the lecture notes he took in the course of Professor R.B. Dingle at the University of Western Australia in 1956, and realized that this course covered in a clear and logically arranged way the vital basic points of the subject and included a large number of illustrating examples and exercises (in the present text most of the examples are with solutions). The following text is a presentation of the subject arranged along the lines of this course for which the author is indebted to his former teacher who, it may be pointed out here, made significant contributions to the subject, particularly in providing the first valid demonstrations of Bose–Einstein condensation (in an ideal gas). This introduction to Statistical Physics, which took the author's notes of this course as a guideline and employs only wave mechanics instead of full operator quantum mechanics, may be of interest to others who are interested in a *Bachelor Degree course*, or equivalent introductory course to the subject, though, of course, supplemented by some additions (apart from the introduction, particularly the section on Bose–Einstein condensation and the problems without worked solutions) and slight expansion throughout. The reader may ask, however, how this text — which is meant to be a first introduction to the subject — differs from other texts on the subject. One answer may be that the central issues here are — clearly separated from applications — classical versus quantum physics, *a priori* probability and degeneracy, distinguishability and indistinguishability, differences between conserved and nonconserved elements, differences in counting of arrangements in the various statistics, and maximization versus averaging of these. In particular, the text proceeds stepwise to the ultimate Darwin–Fowler method of mean values which not only yields exact results but also provides the basis for the rigorous proof of Bose–Einstein condensation as given by Dingle. Applications are mostly relegated to examples. It will be evident from the text that the author also consulted a number of modern and recent monographs on the subject, and compared the treatment in these with that here, also to provide references with further details. For the convenience of students, calculations are generally given in detail.

Harald J.W. Müller–Kirsten

Chapter 1

Introduction

1.1 Introductory Remarks

Statistical physics embraces in particular statistical mechanics and statistical thermodynamics. The word statistical implies already that the subject deals with a large number of elements like the number of atoms or molecules in a macroscopic body. A theoretical treatment of statistical physics thus attempts to describe macroscopic phenomena in terms of microscopic processes. Elementary microscopic processes are obviously those of atoms and molecules. Statistical thermodynamics considers primarily microscopic processes in some enclosure like a box which imposes stringent boundary conditions on the dynamics of the particles. Consequently in a first approach to the subject one considers cases in which the interaction between individual atoms or molecules is of secondary importance, as in the case of a dilute gas in some container. The realization that the motion of atoms, *i.e.* their kinetic energy, is related to the macroscopically observed temperature was a considerable step forward in our understanding of the relationship between atomic physics and classical thermodynamics. Thus a perfect gas suggests itself naturally as a first object to consider, and then the question whether other cases, *e.g.* conduction electrons in a metal, can be considered similarly. And how are solids to be treated in this context? The naive picture of a solid as a lattice with atoms located at lattice sites suggests to consider these in analogy with harmonic oscillators, for instance, one oscillator at every lattice site. The simple one-dimensional harmonic oscillator serves in view of its mathematical simplicity in many areas of physics as a convenient first modelling example. Thus in a first attempt it is an obvious idea to abstract the atoms at lattice sites to such harmonic oscillators whose oscillations describe the vibrations of an atom or molecule. The harmonic oscillator played a vital role in the development of quantized statistics as conceived by Planck: It was Planck's idea of considering the simple harmonic oscillator as statistically equivalent to a normal mode of vibration which led him to discretized energies. Thus both free particles and oscillators play a dominant role in our introduction to the subject here. However, the quantized simple one-dimensional harmonic oscillator has one limitation: Its eigenvalues are nondegenerate. Since degeneracy will be seen to be a characteristic of many particle states of statistical physics it is expedient to consider oscillators also in higher dimensions, for instance, in a model of solids. Our presentation here begins with elementary kinetic theory. We then introduce the concept of *a priori* probability and show that this can be identified with the degeneracy of states. In classical Maxwell–Boltzmann statistics we consider the number of arrangements W of particles among states of various degeneracies and then determine that particular arrangement which appears with maximum probability. We proceed similarly with quantum statistics, taking into account the indistinguishability of elements, the number of elements permissible per state, and whether the elements are conserved or nonconserved, and thus arrive at Bose–Einstein and Fermi–Dirac statistics. In the last chapter we consider the Darwin–Fowler method of mean values and observe that the more rigorously derived results of this method are the same as those obtained with maximization. However, before we consider quantum statistics we introduce the concept of entropy S as defined by Boltzmann by the relation $S = k \ln W$, k being Boltzmann's constant. Since our aim here is a presentation of the basic principles of statistical physics, including a clear distinction between classical and quantum statistics, we assume here a knowledge of basic quantum mechanics as well as elementary thermodynamics such as one acquires in introductory physics courses. Since we make here particular use of thermodynamics, we recapitulate below some of the basic equations of classical thermodynamics with particular reference to the various thermodynamic potentials. We emphasize: This is only a brief summary of selected equations without any attempt to enter into detailed explanations. For details we refer to the book of H.B. Callen [10]. Other texts which devote more space to fundamentals, thermodynamics and applications are *e.g.* those of D. Chandler [12] and D.V. Schroeder [66].

1.2   Thermodynamic Potentials

We summarize here some very general relations of the thermostatics of homogeneous systems. These relations involve quantities known as *thermodynamic potentials*. We assume a constant number of particles N. The volume V is an external parameter. The *second law of thermodynamics* expressed in terms of entropy S, temperature T, heat Q, as well as (mean) internal energy E is generally written in infinitesimal form as

$$TdS = dQ = dE + PdV \tag{1.1}$$

(the first law being the law of conservation of energy, $dQ = dE + PdV$). In this form the energy E is a function of S and V, *i.e.* $E = E(S, V)$. However, one can choose different combinations of independent macroscopic parameters like in mechanics one can derive Newton's equation from the Lagrangian $L(q, \dot{q})$ or from the Hamiltonian $H(q, p)$, the two quantities being related by a *Legendre transform*, *i.e.* the relation $H(q, p) = \dot{q}p - L(q, \dot{q})$, so that $p = \partial L / \partial \dot{q}$ and $\partial H / \partial \dot{q} = 0$. The various forms of thermodynamic potentials are related in a similar way. We can choose as independent macroscopic parameters the following sets:

$$
\begin{array}{ll}
(a) & N, S, V, \\
(b) & N, S, P, \\
(c) & N, T, V, \\
(d) & N, T, P.
\end{array}
$$

In the case of (a) we have $E = E(S, V)$. In order to transform to set (b) we define (in analogy to mechanics above) the quantity $H(S, P)$ called *enthalpy* or *total heat* by the relation

$$H(S, P) := VP + E(S, V), \tag{1.2}$$

which is such that

$$\frac{\partial H}{\partial V} = 0 \quad \text{with} \quad P = -\frac{\partial E}{\partial V}. \tag{1.3}$$

In classical mechanics we have Hamilton's equations

$$\dot{q} = \frac{\partial H(q, p)}{\partial p}, \qquad \dot{p} = -\frac{\partial H(q, p)}{\partial q}. \tag{1.4}$$

Correspondingly we have in the present case:

$$V = \left(\frac{\partial H}{\partial P}\right)_S, \qquad T = \left(\frac{\partial E}{\partial S}\right)_V = \left(\frac{\partial H}{\partial S}\right)_P \tag{1.5}$$

(we shall see that these can also be read off the infinitesimal relations given below). Corresponding to each of the four transformations

$$(S, V) \to (S, P), \quad (S, V) \to (T, V), \quad (T, V) \to (T, P), \quad (S, P) \to (T, P),$$

we require one quantity corresponding to the Hamiltonian in mechanics, and hence four such quantities altogether if we include E. These four quantities, including E are referred to as *thermodynamic potentials*. These are:

$$E(S,V) \qquad \text{called} \quad \textit{``internal energy''},$$

$$H(S,P) = E(S,V) + PV \qquad \text{called} \quad \textit{``enthalpy''} \text{ or}$$
$$(= G(T,P) + TS) \qquad\qquad\qquad \textit{``total heat''},$$

$$F(T,V) = E(S,V) - TS \qquad \text{called} \quad \textit{``free energy''},$$

$$G(T,P) = E(S,V) - TS + PV \quad \text{called} \quad \textit{``free enthalpy''} \text{ or}$$
$$(= F(T,V) + PV) \qquad\qquad\qquad \textit{``Gibbs function''}.$$

We consider in each of these cases the relations corresponding to Hamilton's equations in mechanics.

(a) $E(S,V)$: We have from the second law of thermodynamics

$$dE = TdS - PdV$$

and

$$\therefore \qquad T = \left(\frac{\partial E}{\partial S}\right)_V, \qquad -P = \left(\frac{\partial E}{\partial V}\right)_S, \tag{1.6}$$

and hence the *first Maxwell relation*:

$$\left(\frac{\partial T}{\partial V}\right)_S = -\left(\frac{\partial P}{\partial S}\right)_V. \tag{1.7}$$

(b) $H(S,P)$: We obtain from the definition of H:

$$dH(S,P) = dE + PdV + VdP = TdS + VdP, \tag{1.8}$$

and (recall also that $H(S,P) = E(S,V) + PV$)

$$\therefore \qquad \left(\frac{\partial H}{\partial P}\right)_S = V, \qquad \left(\frac{\partial H}{\partial S}\right)_P = \left(\frac{\partial E}{\partial S}\right)_P = T, \tag{1.9}$$

and hence the *second Maxwell relation*:

$$\left(\frac{\partial V}{\partial S}\right)_P = \left(\frac{\partial T}{\partial P}\right)_S. \tag{1.10}$$

(c) $F(T,P)$: We obtain from the definition of F:

$$dF(T,V) = dE - TdS - SdT = -SdT - PdV, \tag{1.11}$$

and

$$\therefore \quad \left(\frac{\partial F}{\partial T}\right)_V = -S, \quad \left(\frac{\partial F}{\partial V}\right)_T = \left(\frac{\partial E}{\partial V}\right)_T = -P, \quad (1.12)$$

and hence the *third Maxwell relation*:

$$\left(\frac{\partial S}{\partial V}\right)_T = \left(\frac{\partial P}{\partial T}\right)_V. \quad (1.13)$$

(d) $G(T, P)$: We obtain from the definition of G:

$$dG(T, P) = dE - TdS - SdT + PdV + VdP = VdP - SdT, \quad (1.14)$$

and

$$\therefore \quad \left(\frac{\partial G}{\partial T}\right)_P = -S = \left(\frac{\partial F}{\partial T}\right)_V, \quad \left(\frac{\partial G}{\partial P}\right)_T = V, \quad (1.15)$$

and hence the *fourth Maxwell relation*:

$$\left(\frac{\partial S}{\partial P}\right)_T = -\left(\frac{\partial V}{\partial T}\right)_P. \quad (1.16)$$

These results can be read off the infinitesimal relations

$$\left. \begin{aligned} dH &= TdS + VdP, \\ dF &= -SdT - PdV, \\ dG &= S(-dT) + VdP. \end{aligned} \right\} \quad (1.17)$$

We shall see later in statistical thermodynamics that most of the thermodynamical relations required can be derived from the free energy F, which is therefore of particular importance. The natural question to follow is, of course: What is the use of the Maxwell relations? Since our main topic here is not classical thermodynamics, we only illustrate the use of a Maxwell relation below in the consideration of heat capacities. A few useful tricks in manipulating the relations in specific calculations are the following:

$$\left(\frac{\partial x}{\partial y}\right)_z = \left[\left(\frac{\partial y}{\partial x}\right)_z\right]^{-1}, \quad \left(\frac{\partial x}{\partial y}\right)_z = \left(\frac{\partial x}{\partial w}\right)_z \left[\left(\frac{\partial y}{\partial w}\right)_z\right]^{-1}. \quad (1.18)$$

Moreover, from $z = z(x, y)$ we obtain

$$dz = \left(\frac{\partial z}{\partial x}\right)_y dx + \left(\frac{\partial z}{\partial y}\right)_x dy, \quad (1.19)$$

so that for $z = $ const.

$$0 = \left(\frac{\partial z}{\partial x}\right)_y \left(\frac{\partial x}{\partial y}\right)_z + \left(\frac{\partial z}{\partial y}\right)_x \quad \text{or} \quad \left(\frac{\partial x}{\partial y}\right)_z = -\frac{(\partial z/\partial y)_x}{(\partial z/\partial x)_y}. \quad (1.20)$$

1.3   Capacity of Heat

Throughout the text we shall frequently consider a quantity C known as the *capacity of heat* or *specific heat* (the latter frequently being defined as C per particle). This quantity is a measure of the capability of a substance to store heat; hence its name. One defines specifically and very generally the two quantities

$$C_V = \left(\frac{dQ}{dT}\right)_V = T\left(\frac{\partial S}{\partial T}\right)_V,$$

$$C_P = \left(\frac{dQ}{dT}\right)_P = T\left(\frac{\partial S}{\partial T}\right)_P. \tag{1.21}$$

Temperature T and pressure P are easiest to measure. Thus consider for instance $S = S(T, P)$. Then

$$\begin{aligned} dQ &= TdS(T,P) \\ &= T\left[\left(\frac{\partial S}{\partial T}\right)_P dT + \left(\frac{\partial S}{\partial P}\right)_T dP\right] \\ &= C_P dT + T\left(\frac{\partial S}{\partial P}\right)_T dP. \end{aligned} \tag{1.22}$$

But $P = P(V, T)$, so that

$$dP(V,T) = \left(\frac{\partial P}{\partial T}\right)_V dT + \left(\frac{\partial P}{\partial V}\right)_T dV. \tag{1.23}$$

It follows that

$$\frac{dQ}{dT} = C_P + T\left(\frac{\partial S}{\partial P}\right)_T \left\{\left(\frac{\partial P}{\partial T}\right)_V + \left(\frac{\partial P}{\partial V}\right)_T \frac{dV}{dT}\right\}, \tag{1.24}$$

and therefore

$$C_V = \left(\frac{dQ}{dT}\right)_V = C_P + T\left(\frac{\partial S}{\partial P}\right)_T \left(\frac{\partial P}{\partial T}\right)_V. \tag{1.25}$$

With the help of the fourth Maxwell relation (1.16) this equation becomes

$$C_V = C_P - T\left(\frac{\partial V}{\partial T}\right)_P \left(\frac{\partial P}{\partial T}\right)_V. \tag{1.26}$$

One now defines as *expansion coefficient* α the expression

$$\alpha := \frac{1}{V}\left(\frac{\partial V}{\partial T}\right)_P. \tag{1.27}$$

One also defines as *isothermal compressibility* κ_T and as *adiabatic compressibility* κ_S the expressions[*]

$$\kappa_T := -\frac{1}{V}\left(\frac{\partial V}{\partial P}\right)_T, \quad \kappa_S := -\frac{1}{V}\left(\frac{\partial V}{\partial P}\right)_S, \tag{1.28}$$

and as *stress coefficient* β:

$$\beta := \frac{1}{P}\left(\frac{\partial P}{\partial T}\right)_V. \tag{1.29}$$

These parameters are introduced because they are relatively easy to determine experimentally. We can now express the difference $C_V - C_P$ in terms of these parameters. Thus, using Eq. (1.20), we have

$$\left(\frac{\partial P}{\partial T}\right)_V = -\frac{(\partial V/\partial T)_P}{(\partial V/\partial P)_T} = \frac{\alpha V}{\kappa V} = \frac{\alpha}{\kappa}. \tag{1.30}$$

It follows that

$$C_V = C_P - \frac{\alpha^2 V T}{\kappa}. \tag{1.31}$$

Concerning further properties of heat capacities, in particular their behaviour in approaching the absolute zero of temperature (and the third law of thermodynamics according to which $S(T,V)$ and $S(T,P)$ approach a constant independent of external parameters) we refer to the book of Callen [10]. The same applies to the question of which experimental data are required in order to determine S, E, F. Throughout we consider only equilibrium statistical mechanics.

1.4   Frequently Used Terms

In closing this recapitulation of some aspects of classical thermodynamics we summarize some frequently used terms. We mentioned already W, the number of arrangements of (say N) elements (atoms, molecules). It is this quantity which Boltzmann related to entropy in his famous formula $S = k \ln W$ (the subject of Chapter 5). Considering dS, we have in general (for external parameters X_i like V or a magnetic field H)

$$dS = k\, d\ln W(E, X_i, t) \tag{1.32}$$

together with the relations

$$dQ = dE + dW \quad \text{and} \quad dW = \sum_i \phi_i dX_i \overset{(1.1)}{=} PdV + MdH. \tag{1.33}$$

---

[*]See *e.g.* A.D. Buckingham [8], pp.104, 194.

Note that t is here a macroscopic time. Then:

A process is *quasistatic* if

$$\frac{\partial W}{\partial t} \sim 0. \qquad (1.34)$$

A process is *reversible* if

$$dS = 0 \qquad (W_{\text{initial}} = W_{\text{final}}). \qquad (1.35)$$

A process is said to be *isothermal* if

$$dT = 0 \qquad (T_{\text{initial}} = T_{\text{final}}). \qquad (1.36)$$

A process is said to be *adiabatic* (no heat added or subtracted or arbitrarily close to equilibrium) if

$$dQ = 0. \qquad (1.37)$$

1.5   Applications and Examples

Example 1.1: Equivalence of E and entropy S equilibrium principles
Assuming E is the total energy of a system A in contact with a heat bath R, and S the entropy, show that the equilibrium values of the internal parameters (*e.g.* temperature T) can be determined from either the *minimum energy principle* or the *maximum entropy principle*, *i.e.*

$$\delta E = 0 \quad \text{with} \quad \delta^2 E > 0, \qquad \text{or} \qquad \delta S = 0 \quad \text{with} \quad \delta^2 S < 0. \qquad (1.38)$$

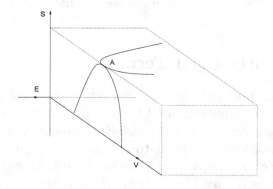

Fig. 1.1 Maximum S, minimum E of equilibrium state A.

Solution: We establish the equivalence of the two principles by showing that if the energy is not minimal, then at equilibrium the entropy can not be maximal. Thus we begin by assuming that at a given maximal entropy the energy E is not minimal. We consider the *second law of thermodynamics*,

$$TdS = dQ = dE + dW. \qquad (1.39)$$

We keep the entropy constant, so that $dQ = 0$. Since the energy is not that of a minimum, we can lower the energy by an amount dE by allowing the system to perform the amount of work dW

(with no change in the entropy — for instance by allowing a piston to slowly push the gas outside, quasistatically and adiabatically). Thereafter we re-establish the original energy of the system by adding the amount of heat dQ. Then the energy of the system is the same as before. However, we have increased the entropy by the amount dQ/T. Also we assumed that the system occupied a state of maximal entropy in its state of equilibrium. Hence our assumption must be wrong, *i.e.* the energy of the system in equilibrium at maximum entropy must be minimal. Hence the equilibrium state follows from both principles. This is what one would expect, as a state of equilibrium is usually connected with a minimum of energy. This equivalence is illustrated in Fig. 1.1 in which the point A represents a state of equilibrium with in one case S maximal at $E = $ const., and in the other case E minimal at $S = $ const.

Example 1.2: Heat capacity from equation of state
Show that $C_V(T, V)$ can be derived from the equation of state.

Solution: We choose T and V as independent parameters of S, *i.e.* we start from $S = S(T, V)$. Then

$$dS = \left(\frac{\partial S}{\partial T}\right)_V dT + \left(\frac{\partial S}{\partial V}\right)_T dV \overset{(1.21),(1.13)}{=} \frac{1}{T}C_V\, dT + \left(\frac{\partial P}{\partial T}\right)_V dV. \qquad (1.40)$$

Thus

$$C_V = T\left(\frac{\partial S}{\partial T}\right)_V, \qquad (1.41)$$

and

$$
\begin{aligned}
\left(\frac{\partial}{\partial V}C_V\right)_T &= \left(\frac{\partial}{\partial V}\right)_T \left[T\left(\frac{\partial S}{\partial T}\right)_V\right] \\
&= T\left(\frac{\partial}{\partial T}\right)_V\left(\frac{\partial S}{\partial V}\right)_T \\
&\overset{(1.13)}{=} T\left(\frac{\partial}{\partial T}\right)\left(\frac{\partial P}{\partial T}\right)_V,
\end{aligned}
\qquad (1.42)
$$

i.e.

$$\left(\frac{\partial C_V}{\partial V}\right)_T = T\left(\frac{\partial^2 P}{\partial T^2}\right)_V. \qquad (1.43)$$

We can obtain the right hand side from the equation of state $(P = \cdots)$. Hence C_V follows from the equation

$$C_V(T, V) = C_V(T, V_1) + \int_{V_1}^{V} \left(\frac{\partial C_V(T, V')}{\partial V'}\right)_T dV'. \qquad (1.44)$$

1.6   Problems without Worked Solutions

Example 1.3: Van-der-Waals gas
The equation of state of a van-der-Waals gas is[†]

$$\left(P + \frac{a^2}{V^2}\right)(V - b) = NkT, \quad a, b > 0. \qquad (1.45)$$

Derive for this gas an expression for the difference $C_P - C_V$, and from this its value in the limit $V \to \infty$. (Answer for $V \to \infty$: $C_P - C_V = Nk$).

[†]Note that for a perfect gas $PV = (N/N_0)RT = NkT$. Here k is Boltzmann's constant given by $k = R/N_0$, where N_0 is Avogadro's number (the number of particles of mass m of an element of atomic weight $N_0 m$), and R is the gas constant, and N the number of particles.

Example 1.4: Velocity of sound in a gas

The velocity of sound in a gas is given by the relation[‡]

$$v_s = \frac{1}{\sqrt{\rho \kappa_S}} = \sqrt{-\frac{V}{\rho}\left(\frac{\partial P}{\partial V}\right)_S}, \tag{1.46}$$

where ρ is the density of the gas. The capacity of heat at constant pressure can be expressed as

$$C_P(T) = A + BT + CT^2, \tag{1.47}$$

and A, B, C can be looked up in Tables. Express v_s in terms of P, T and C_P for a perfect gas with equation of state $PV = RT$ for one mole and $R = 8.31$ J/mol.K (*i.e.* $\rho v_s = m$ and the molar mass $M = m/n$, where n, the number of moles, can be looked up in Tables). (Answer: $v_s = \sqrt{PC_P(T)/\rho\{C_P(T) - R\}}$).

Example 1.5: The adiabatic index of air

The equation of state for a gas which does not obey the ideal gas law is $PV^\gamma = $ constant, where γ is the so-called adiabatic index. Show that the equation for the speed of sound in this case is given by (m being the mass of one particle of the gas)

$$v_s = \sqrt{\frac{\gamma RT}{m}} \tag{1.48}$$

for one mole. The speed of sound in air is approximately 350 m/s. What is the value of γ? (Answer: 1.4).

Example 1.6: The gas law and causality

What does causality imply for the gas law? (Answer: $v_s < c$, velocity of light).

[‡]The derivation of this formula can be found *e.g.* in E.G. Richardson [61], Chapter 1, or in S.G. Starling and A.J. Woodall [72], pp.823-828. But observe the difference between the two coefficients defined in Eq. (1.28) above.

Chapter 2

Statistical Mechanics of an Ideal Gas (Maxwell)

2.1 Introductory Remarks

A convenient way to begin is by considering elementary aspects of the *kinetic theory* of gases* as developed by Maxwell and others since this leads immediately to the Maxwell–Boltzmann exponential factor (the Maxwell distribution involving this factor only with kinetic energy, hence kinetic theory), to integrals over exponential functions which will occur frequently throughout the text, and in addition enables us to introduce the *method of Lagrangian multipliers*, which is required for the enforcement of subsidiary conditions. A convenient example throughout is the case of a perfect or ideal gas.

2.2 Maxwell's Treatment

We consider the molecules of a gas, and consider these as moving around in a container independently of one another and with various speeds, as depicted in Fig. 2.1. Consider one direction, like that of coordinate x.

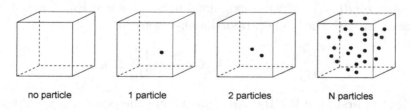

no particle   1 particle   2 particles   N particles

Fig. 2.1 Molecules of a gas as classical particles.

*For a somewhat different treatment see F. Reif [59], Sec. 7.9.

Let u be the velocity of a molecule in the direction of x. Define $\Pi_u du$ to be the probability of a molecule of a gas having a velocity in the given x-direction between u and $u + du$. Let the molecule have similarly velocity components between v and $v + dv$ in the y-direction and velocity components between w and $w + dw$ in the z-direction. The three directions are independent of one another. Hence the probability for these three directions simultaneously is the product

$$\Pi_{uvw} du dv dw = \Pi_u \Pi_v \Pi_w du dv dw. \tag{2.1}$$

We consider molecules which all have the same constant speed c. Then

$$c^2 = u^2 + v^2 + w^2. \tag{2.2}$$

Differentiating the last equation we obtain

$$0 = c\, dc = u\, du + v\, dv + w\, dw. \tag{2.3}$$

Since the distribution is assumed to be *isotropic*, all directions are equally likely. Therefore Π_{uvw} has to be the same for all (u, v, w) which correspond to the same speed c. It follows therefore that

$$\Pi_{uvw} \equiv \Pi_u \Pi_v \Pi_w = \text{const.} \tag{2.4}$$

Since this implies that $d\Pi_{uvw} = 0$, we obtain by dividing this by $\Pi_u \Pi_v \Pi_w$ (primes denoting differentiation with respect to the indicated variable):

$$0 = \frac{\Pi'_u}{\Pi_u} du + \frac{\Pi'_v}{\Pi_v} dv + \frac{\Pi'_w}{\Pi_w} dw, \tag{2.5}$$

where

$$\Pi'_u = \frac{\partial \Pi_u}{\partial u}, \quad \Pi'_v = \frac{\partial \Pi_v}{\partial v}, \quad \Pi'_w = \frac{\partial \Pi_w}{\partial w}.$$

In order to determine the functions Π_u, Π_v, Π_w, we have to solve Eq. (2.5) subject to the restriction of Eq. (2.3). This is achieved with the method of *Lagrangian multipliers* which is explained in Sec. 2.3 and Example 2.1 below. The result is, with the same proportionality constant,

$$\frac{\Pi'_u}{\Pi_u} \propto u, \quad \frac{\Pi'_v}{\Pi_v} \propto v, \quad \frac{\Pi'_w}{\Pi_w} \propto w. \tag{2.6}$$

Consider the first case $\Pi'_u / \Pi_u \propto u$. Integrating the equation, we obtain

$$\int \frac{d\Pi_u}{\Pi_u} \propto \int u\, du, \quad \ln \Pi_u \propto u^2 + \text{const.} \tag{2.7}$$

or

$$\Pi_u = ae^{-\alpha u^2}, \tag{2.8}$$

where a and α are as yet undetermined constants.

We determine the constants a and α of the *Maxwell distribution* (2.8) as follows. It is a certainty that the molecule has some velocity between $-\infty$ and $+\infty$ (this assumes that the velocity is real). This condition implies the *normalization*

$$\int_{-\infty}^{\infty} \Pi_u du = 1 \quad \rightarrow \quad a \int_{-\infty}^{\infty} e^{-\alpha u^2} du = 1, \quad a\sqrt{\frac{\pi}{\alpha}} = 1, \quad a = \sqrt{\frac{\alpha}{\pi}}, \tag{2.9}$$

where we used a result of Example 2.1 for the evaluation of the Gaussian integral. Hence

$$\Pi_u = \sqrt{\frac{\alpha}{\pi}} e^{-\alpha u^2}. \tag{2.10}$$

Throughout the entire text we require the evaluation of integrals over these Gaussian exponentials. Therefore we deal with these integrals immediately and refer back to the results in the text.

Example 2.1: Evaluation of Gaussian integrals
Evaluate (for use later in this text) the Gaussian integrals

$$I_n(\alpha) := \int_0^{\infty} u^n e^{-\alpha u^2} du, \quad n = 0, 1, 2, 3, 4, \ldots. \tag{2.11}$$

Solution: One sets $v = \alpha u^2$, so that by the definition of the *gamma function* $\Gamma(x+1) = x!$ (for convenience we use here the factorial notation)

$$I_n(\alpha) = \frac{1}{2\alpha^{(n+1)/2}} \int_0^{\infty} v^{(n-1)/2} e^{-v} dv = \frac{[(n-1)/2]!}{2\alpha^{(n+1)/2}}. \tag{2.12}$$

We distinguish between odd and even values of n.

1. *n odd*: In this case $(n-1)/2$ is an integer and its factorial follows at once with $p! = p(p-1)(p-2)\cdots 1$, when p is an integer.

2. *n even*: It is convenient to evaluate first the integral for $n = 0$. From Eq. (2.11):

$$I_0(\alpha) = \int_0^{\infty} e^{-\alpha u^2} du = \frac{[-1/2]!}{2\sqrt{\alpha}}. \tag{2.13}$$

Knowing this result, $I_n(\alpha)$ for general *even n* can be deduced by either of the following two methods:

(a) We can reduce $I_n(\alpha)$ to the form of $I_0(\alpha)$ with the help of the recurrence relation $p! = p(p-1)!$. Thus, for instance,

$$I_2(\alpha) = \frac{1}{2\alpha^{3/2}} \left[\frac{1}{2}\right]! = \frac{1}{4\alpha^{3/2}} \left[-\frac{1}{2}\right]!. \tag{2.14}$$

(b) Alternatively we can differentiate with respect to α under the integral sign. Thus

$$I_n(\alpha) = \int_0^\infty u^n e^{-\alpha u^2} du = -\frac{\partial}{\partial \alpha} \int_0^\infty u^{n-2} e^{-\alpha u^2} du = -\frac{\partial}{\partial \alpha} I_{n-2}(\alpha). \qquad (2.15)$$

For instance,

$$I_2(\alpha) = -\frac{\partial}{\partial \alpha} I_0(\alpha) \overset{(2.13)}{=} -\frac{1}{2}\left[-\frac{1}{2}\right]! \frac{\partial}{\partial \alpha}(\alpha^{-1/2}) = \frac{1}{4\alpha^{3/2}}\left[-\frac{1}{2}\right]!. \qquad (2.16)$$

Next we evaluate $I_0(\alpha)$ and hence $[-1/2]!$. Consider

$$I_0^2(\alpha) = \int_0^\infty e^{-\alpha u^2} du \int_0^\infty e^{-\alpha v^2} dv = \int_0^\infty \int_0^\infty e^{-\alpha(u^2+v^2)} du dv. \qquad (2.17)$$

Now introduce the plane polar coordinates r, θ, with $u = r\cos\theta, v = r\sin\theta, u^2 + v^2 = r^2, du dv = r dr d\theta$. Then, with $s = \alpha r^2$,

$$I_0^2(\alpha) = \int_{\theta=0}^{\pi/2} d\theta \int_{r=0}^\infty e^{-\alpha r^2} r dr = \frac{\pi}{2} \frac{1}{2\alpha} \int_0^\infty e^{-s} ds = \frac{\pi}{4\alpha}. \qquad (2.18)$$

Hence

$$I_0(\alpha) = \frac{1}{2\alpha^{1/2}} \pi^{1/2}, \quad \text{and} \quad \left[-\frac{1}{2}\right]! = \pi^{1/2}. \qquad (2.19)$$

One can now evaluate in particular the following integrals:

$$\int_0^\infty u e^{-\alpha u^2} du = \frac{1}{2\alpha}, \qquad \int_0^\infty e^{-\alpha u^2} du = \frac{1}{2}\sqrt{\frac{\pi}{\alpha}},$$

$$\int_0^\infty u^3 e^{-\alpha u^2} du = \frac{1}{2\alpha^2}, \qquad \int_0^\infty u^2 e^{-\alpha u^2} du = \frac{1}{4}\sqrt{\frac{\pi}{\alpha^3}},$$

$$\int_0^\infty u^5 e^{-\alpha u^2} du = \frac{1}{\alpha^3}, \qquad \int_0^\infty u^4 e^{-\alpha u^2} du = \frac{3}{8}\sqrt{\frac{\pi}{\alpha^5}},$$

$$\int_0^\infty u^7 e^{-\alpha u^2} du = \frac{3}{\alpha^4}. \qquad \int_0^\infty u^n e^{-\alpha u^2} du = \frac{1}{2}\left(\frac{n-1}{2}\right)! \alpha^{-(n+1)/2}. \qquad (2.20)$$

Various of these integrals will be required in this text.

2.3   Lagrange's Method of Multipliers

The essential enforcement of subsidiary conditions (like constraints) with the help of Lagrange multipliers is a basic problem of statistical theory, and therefore deserves to be considered at an early stage. We illustrate the *method of Lagrange multipliers*, also called the *method of undetermined multipliers,*[†]

[†] The presentation here is similar to that in F. Reif [59], Appendix 10, and in J.E. Mayer and M.G. Mayer [45], pp.433-435. On a still simpler level it is used by R.P.H. Gasser and W.G. Richards [28], pp.15-16. The reader will recall Lagrange multipliers from classical mechanics, as parameters multiplying the forces of constraints. This connection is explained by S.-K. Ma [42], pp.88-94.

by an example which can be modified appropriately to apply in many other cases. Thus, we assume we have the following equation

$$\sum_{i=1}^{I} f_i(n_i) dn_i = 0, \tag{2.21}$$

which has to be solved for n_i subject to the one condition

$$\sum_{i=1}^{I} n_i = N \text{ (const.)}, \quad i.e. \quad \sum_{i=1}^{I} dn_i = 0. \tag{2.22}$$

Since the latter is one condition on otherwise I different n_i, $I-1$ of the n_i are independent. We multiply Eq. (2.22) by a quantity λ which is independent of i and will be determined later, and we add this equation to Eq. (2.21). Then

$$\sum_{i=1}^{I} [f_i(n_i) + \lambda] dn_i = 0. \tag{2.23}$$

Let's now suppose that λ be determined from the equation

$$f_1(n_1) + \lambda = 0 \quad \text{for any } n_1. \tag{2.24}$$

With this condition Eq. (2.23) reduces to

$$\sum_{i=2}^{I} [f_i(n_i) + \lambda] dn_i = 0. \tag{2.25}$$

But by Eq. (2.22) all the remaining dn_i's $(dn_2, dn_3, \ldots, dn_I)$ are now independent, and determine the resultant value of

$$dn_1 = -\sum_{i=2}^{I} dn_i,$$

which does not appear in Eq. (2.25). The only way Eq. (2.25) can be satisfied for arbitrary independent variations $dn_2, dn_3, \ldots, dn_I$ is by setting

$$f_i(n_i) + \lambda = 0 \quad \text{for} \quad i = 2, \ldots, I. \tag{2.26}$$

Combining Eqs. (2.24) and (2.26), we obtain

$$f_i(n_i) + \lambda = 0 \quad \text{for all } i = 1, \ldots, I. \tag{2.27}$$

Rewriting Eq. (2.27) in terms of the inverse f^{-1} of f, the multiplier λ is now determined by Eq. (2.22) *i.e.* by

$$N = \sum_{i=1}^{I} n_i = \sum_{i=1}^{I} f_i^{-1}(-\lambda). \tag{2.28}$$

In Sec. 2.2 we have instead of Eqs. (2.21) and (2.22), the equations (2.5) and (2.3), *i.e.*

$$\sum_{i=1}^{3} f_i(n_i)dn_i = 0, \qquad \sum_{i=1}^{3} n_idn_i = 0, \tag{2.29}$$

where $n_1 = u, n_2 = v, n_3 = w$ and

$$f_1(n_1) = \frac{\Pi'_u}{\Pi_u}, \quad f_2(n_2) = \frac{\Pi'_v}{\Pi_v}, \quad f_3(n_3) = \frac{\Pi'_w}{\Pi_w}. \tag{2.30}$$

Example 2.2: Lagrange multiplier and Maxwell distribution

Use the method of Lagrange multipliers to derive the proportionalities (2.6), *i.e.* $\Pi'_i / \Pi_i \propto n_i$.

Solution: Corresponding to Eq. (2.3) we have, with $n_i \to u, v, w$,

$$\sum_i n_idn_i = 0, \tag{2.31}$$

and corresponding to Eq. (2.5), with $f_i(n_i) \to \Pi'_u / \Pi_u, \Pi'_v / \Pi_v, \Pi'_w / \Pi_w$, we have

$$\sum_i f_i(n_i)dn_i = 0. \tag{2.32}$$

Proceeding as above, we obtain with a Lagrange multiplier λ the relation

$$\sum_{i=1}^{I}[f_i(n_i) + \lambda n_i]dn_i = 0. \tag{2.33}$$

Suppose the parameter λ is determined by the equation

$$f_1(n_1) + \lambda n_1 = 0. \tag{2.34}$$

Then

$$\sum_{i=2}^{I}[f_i(n_i) + \lambda n_i]dn_i = 0. \tag{2.35}$$

Since the $I - 1$ quantities $n_i, i = 2, 3, \ldots, I$, are independent, the only way this relation can be satisfied is by demanding

$$f_i(n_i) + \lambda n_i = 0 \quad \text{for all } i. \tag{2.36}$$

Thus as claimed in Eq. (2.6)

$$f_i \equiv \frac{\Pi'_i}{\Pi_i} \propto n_i. \tag{2.37}$$

2.4 Applications

We consider three applications of the above *Maxwell distribution function*.

2.4.1 Pressure exerted on the wall of a vessel

As a first application we consider the pressure exerted by a gas on the wall of a vessel.[‡] We take the wall perpendicular to the direction of the velocity u, as indicated in Fig. 2.2.

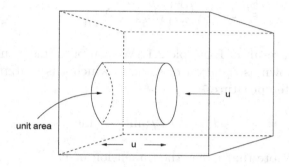

Fig. 2.2 Molecules of velocity u impinging on unit area of a wall.

Let N be the total number of molecules per unit volume. Then the number of molecules having a velocity between u and $u + du$ hitting unit area of the wall per unit time is

$$N\Pi_u du \times u \times 1.$$

Let us understand the meaning of this by rewriting the expression with the help of Eq. (2.9) as

$$N \times (1 \times u) \times \frac{\Pi_u\, du}{\int_{-\infty}^{\infty} \Pi_u\, du}. \tag{2.38}$$

Here $N \times (1 \times u)$ is the number of molecules of velocity u or between u and $u+du$ impinging in one second (*i.e.* per second) on unit area of the wall. But each such molecule has this velocity u in $(u, u + du)$ only with probability $\Pi_u\, du / \int_{-\infty}^{\infty} \Pi_u\, du = \Pi_u\, du$. The momentum change of each molecule of mass m reflected at the wall is $2mu$. It follows that the (probabilistic) pressure P of the molecules on the wall is:[§]

[‡]See also F. Reif [59], Sec. 7.11.

[§]Here we deal with the pressure P due to the force F of an instantaneous impulse at a time t_0 defined by $\lim_{\tau \to 0} \int_{t_0}^{t_0+\tau} F d\tau = \lim_{\tau \to 0} \int_{t_0}^{t_0+\tau} \frac{dp_x}{dt} dt = [p_x]_0^\tau = 2p_x$.

$$\text{pressure } P = \frac{\text{force}}{\text{area}} = \frac{\text{rate of change of momentum}}{\text{area}}$$

$$= \int_0^\infty N\Pi_u du \times u \times 2mu, \tag{2.39}$$

where the lower integration limit is zero since only molecules initially moving *towards* the wall are to be considered (the others do not contribute). Then

$$\text{pressure } P \;=\; 2mN \int_0^\infty \Pi_u u^2 du = 2mN\sqrt{\frac{\alpha}{\pi}} \int_0^\infty u^2 e^{-\alpha u^2} du$$

$$\overset{(2.20)}{=} 2mN\sqrt{\frac{\alpha}{\pi}}\frac{1}{4}\sqrt{\frac{\pi}{\alpha^3}} = N\frac{m}{2\alpha}, \tag{2.40}$$

where we used a result of Example 2.1. We compare this result now with the *gas law*, also known as *equation of state*, in which k is Boltzmann's constant and T denotes temperature,[¶]

$$P = NkT/\text{unit volume}. \tag{2.41}$$

It follows that (note that α has the dimension of $1/u^2$)

$$\frac{m}{2\alpha} = kT, \qquad \alpha = \frac{m}{2kT}.$$

Thus finally we have (*cf.* Eq. (2.10))

$$\Pi_u = \sqrt{\frac{m}{2\pi kT}} e^{-mu^2/2kT}. \tag{2.42}$$

We note an important point. By appealing to the gas law we have introduced the concept of *temperature* purely phenomenologically at this stage. But we shall identify this quantity T later with that defined statistically by Eq. (5.18).

2.4.2   Effusion of gas through a hole

As a second example we consider the case of gas leaking out of the hole of a vessel[‖] as depicted in Fig. 2.3.

[¶]This law of a perfect gas (also called ideal gas) is at this point of the text assumed to be known. In the present case it is used for the determination of the constant α in the Maxwell distribution. The law will be derived later in Example 5.5 (see Eqs. (5.57) and (5.60)).

[‖]See also F. Reif [59], Sec. 7.12.

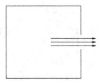

Fig. 2.3 Effusion of gas through a hole in a wall.

Let n_0 be the number of molecules hitting unit area of the container per unit time. Then this is:

$$n_0 = \int_0^\infty u \times \underbrace{N\Pi_u du}_{\text{number of molecules per unit volume with u}\in(\text{u,u+du})}, \qquad (2.43)$$

where the integration begins from $u = 0$ since only molecules moving *towards* unit area are to be considered and velocity $\times$ number/volume = number per unit area per unit time. It follows that (using the value of the integral from Example 2.1)

$$n_0 = N\sqrt{\frac{m}{2\pi kT}} \int_0^\infty u e^{-mu^2/2kT} du \overset{(2.20)}{=} N\sqrt{\frac{m}{2\pi kT}}\frac{kT}{m} = N\sqrt{\frac{kT}{2\pi m}}. \quad (2.44)$$

2.4.3   Thermionic emission

Next we consider a classical treatment of the electrons in a metal as they approach the latter's surface at $x = 0$ to the air above, as indicated in Fig. 2.4.** The number of electrons getting out of the metal is proportional to n_0, and also to the potential between the metal and the air above. It is assumed that an electron can escape from the metal only if it has a velocity component u in the x-direction such that $mu^2/2 \geq$ the energy of escape (not all electrons which can escape do, and therefore one has to allow for this by introducing an efficiency factor). Thus (using Eqs. (2.50b), (2.52b), see Examples 2.3 and 2.5):[††]

$$n_0 = \frac{1}{2}\overline{|u|}N = \frac{1}{4}\bar{c}N, \qquad (2.45)$$

where N is the number of electrons per unit volume in the metal, and $\overline{|u|}$ and $\bar{c}$ are average values of the velocities $|u|$ and c. The classical theory of thermionic emission is also considered in Example 2.8.

[**]The proper nonclassical treatment of the electrons in thermionic emission using Fermi–Dirac statistics is considered later in Examples 10.7 to 10.9.

[††]A derivation is also given in the book of J.H. Jeans [36], p.58.

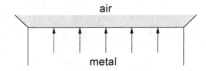

Fig. 2.4 Electrons approaching the surface of a metal.

2.5   Distribution Function for all Directions

So far we applied Maxwell's distribution only in one direction. We now wish to obtain it for all directions. To this end we introduce in the velocity space (u, v, w) spherical coordinates (c, θ, ϕ) as indicated in Fig. 2.5. Then the volume element in velocity space is given by

$$du\,dv\,dw = c^2 dc \sin\theta d\theta d\phi.$$

We have as before:

$$\Pi_{uvw} du\,dv\,dw = \Pi_u \Pi_v \Pi_w du\,dv\,dw \quad \text{with} \quad \Pi_u = \sqrt{\frac{m}{2\pi kT}} e^{-mu^2/2kT}, \ etc.$$
$$(2.46)$$

From these expressions we obtain

$$\Pi_{c\theta\phi} dc\,d\theta\,d\phi = \text{probability that the molecule has speed between}$$
$$c, c + dc \text{ with direction } \theta, \theta + \mathrm{d}\theta, \phi, \phi + \mathrm{d}\phi$$
$$= \left(\frac{m}{2\pi kT}\right)^{3/2} e^{-mc^2/2kT} c^2 dc \sin\theta d\theta d\phi, \qquad (2.47)$$

where $c^2 = u^2 + v^2 + w^2$.

As far as the direction is concerned, the probability is proportional to

$$\frac{\text{solid angle}}{4\pi}.$$

The integration over the angles θ and ϕ is standard, and one obtains, since

$$\int_0^\pi \sin\theta d\theta \int_0^{2\pi} d\phi = 4\pi,$$

as the probability that the molecule has a speed between c and $c + dc$ the result:

$$\Pi_c dc = 4\pi \left(\frac{m}{2\pi kT}\right)^{3/2} e^{-mc^2/2kT} c^2 dc. \qquad (2.48)$$

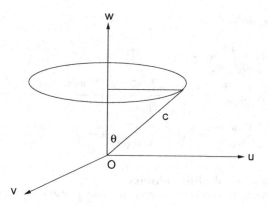

Fig. 2.5 Velocity space spherical coordinates.

2.6   Applications and Examples

Example 2.3: Maxwell's distribution law applied to gases

The probability of a molecule having a velocity between u and $u + du$ in the x-direction is given by the Maxwell distribution (*cf.* Fig. 2.6)

$$\Pi_u du = \sqrt{\frac{m}{2\pi kT}} e^{-mu^2/2kT} du. \qquad (2.49)$$

Verify the following properties:

$$\int_{-\infty}^{\infty} \Pi_u du = 1, \qquad (2.50a)$$

$$\overline{|u|} = \sqrt{\frac{2kT}{\pi m}}, \qquad (2.50b)$$

$$\overline{u^2} = \frac{kT}{m}. \qquad (2.50c)$$

In these expressions the overline denotes the *average* of the quantity concerned. The average of a quantity is defined by the sum (integral) over this quantity weighted by the normalized (here microscopic) energy weighting or probability factor, *i.e.* the Maxwell distribution, the sum extending over all possible values of the weighting factor.

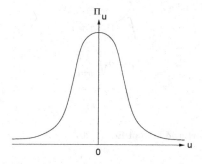

Fig. 2.6 The Maxwell distribution Π_u.

Solution: We use the results (2.20).

$$(a): \int_{-\infty}^{\infty} \Pi_u du = \int_{-\infty}^{\infty} \sqrt{\frac{m}{2\pi kT}} e^{-mu^2/2kT} du = 2\sqrt{\frac{m}{2\pi kT}} \frac{1}{2} \sqrt{\frac{\pi 2kT}{m}} = 1,$$

$$(b): \overline{|u|} = \frac{2\int_0^\infty \Pi_u u\, du}{2\int_0^\infty \Pi_u du} = \frac{\int_0^\infty \sqrt{\frac{m}{2\pi kT}} e^{-mu^2/2kT} u\, du}{\int_0^\infty \sqrt{\frac{m}{2\pi kT}} e^{-mu^2/2kT}\, du} = \frac{2kT}{2m} 2\sqrt{\frac{m}{2kT\pi}} = \sqrt{\frac{2kT}{\pi m}},$$

$$(c): \overline{u^2} = \frac{2\int_0^\infty \Pi_u u^2\, du}{2\int_0^\infty \Pi_u du} = \frac{1}{4}\sqrt{\frac{\pi}{\alpha^3}} \frac{1}{(1/2)\sqrt{\pi/\alpha}} = \frac{1}{2\alpha} = \frac{1}{2}\frac{2kT}{m} = \frac{kT}{m}.$$

Example 2.4: Most probable velocity

Show (using the Maxwell distribution law) that the most probable velocity of a particle is zero. Explain!

Solution: The probability that the particle has a velocity between u and $u + du$ is given by the Maxwell expression, also illustrated in Fig. 2.6,

$$\Pi_u = \sqrt{\frac{m}{2\pi kT}} e^{-mu^2/2kT}.$$

$$\therefore \quad \frac{d\Pi_u}{du} = \sqrt{\frac{m}{2\pi kT}}\left[-\frac{1}{2}m\frac{1}{kT}\right]2u e^{-mu^2/2kT}.$$

This expression vanishes if $u = 0$ or $e^{-mu^2/2kT} = 0$. In the latter case $u = \pm\infty$. We go to the second derivative:

$$\frac{d^2\Pi_u}{du^2} = -\frac{1}{2}m\frac{2}{kT}\sqrt{\frac{m}{2\pi kT}}\left[e^{-mu^2/2kT} - \frac{1}{2}m\frac{1}{kT}2u^2 e^{-mu^2/2kT}\right].$$

This vanishes if $u = \pm\infty$ or $[1 - mu^2/kT] = 0$, *i.e.* $u = \sqrt{kT/m}$, and is negative at $u = 0$. Therefore Π_u is a maximum at $u = 0$. It is a certainty that the particle has some velocity between $-\infty$ and $+\infty$, assuming the veocity is real.

Example 2.5: Probability of a molecule having speed c

Given that the probability of a molecule having a speed between c and $c + dc$ is given by the Maxwell distribution law

$$\Pi_c dc = 4\pi\left(\frac{m}{2\pi kT}\right)^{3/2} e^{-mc^2/2kT} c^2 dc \qquad (2.51)$$

show that:

$$\int_0^\infty \Pi_c dc = 1, \qquad (2.52a)$$

$$\bar{c} = 2\sqrt{\frac{2kT}{\pi m}}, \qquad (2.52b)$$

$$\overline{c^2} = \frac{3kT}{m}. \qquad (2.52c)$$

Solution: We have (again with the help of Eq. (2.20))

$$\int_0^\infty \Pi_c dc = \int_0^\infty 4\pi\left(\frac{m}{2\pi kT}\right)^{3/2} e^{-mc^2/2kT} c^2 dc = 4\pi\left(\frac{m}{2\pi kT}\right)^{3/2} \frac{1}{4}\sqrt{\frac{8\pi k^3 T^3}{m^3}} = 1. \qquad (2.53)$$

Next we recall that the expression

$$\Pi_c dc = 4\pi \left(\frac{m}{2\pi kT}\right)^{3/2} e^{-mc^2/2kT} c^2 dc \tag{2.54}$$

is the fraction of the *number* of molecules having a speed between c and $c + dc$. Hence

$$\bar{c} = \frac{\int_0^\infty 4\pi (m/2\pi kT)^{3/2} e^{-mc^2/2kT} c^2 . c . dc}{\int_0^\infty 4\pi (m/2\pi kT)^{3/2} e^{-mc^2/2kT} c^2 . dc} = \frac{1}{2\alpha^2} \frac{4\sqrt{\alpha^3}}{\sqrt{\pi}} = 2\sqrt{\frac{1}{\alpha\pi}} = 2\sqrt{\frac{2kT}{m\pi}}. \tag{2.55}$$

Finally we consider (again as in Eq. (2.20))

$$\overline{c^2} = \frac{\int_0^\infty \Pi_c c^2 . c^2 dc}{\int_0^\infty \Pi_c c^2 . dc} = \frac{3}{8}\sqrt{\frac{\pi}{\alpha^5}} 4\sqrt{\frac{\alpha^3}{\pi}} = \frac{3}{2}\frac{1}{\alpha} = \frac{3}{2}\frac{2kT}{m} = \frac{3kT}{m}. \tag{2.56}$$

Alternatively we could have argued (*cf.* Sec. 4.3.6):

$$\frac{1}{2}m\overline{c^2} = \frac{3}{2}kT, \quad \therefore \quad \overline{c^2} = \frac{3kT}{m}. \tag{2.57}$$

We see that temperature is an expression of the mean of the square of velocity.

Example 2.6: Most probable speed of a molecule
Show (using the Maxwell distribution law) that the most probable speed of a molecule is given by $\sqrt{2kT/m}$.

Solution: We have

$$\Pi_c = 4\pi \left(\frac{m}{2\pi kT}\right)^{3/2} e^{-mc^2/2kT} c^2. \tag{2.58}$$

Differentiating the expression with respect to c we obtain:

$$\frac{d\Pi_c}{dc} = 4\pi \left(\frac{m}{2\pi kT}\right)^{3/2} \left\{ 2c - \frac{1}{2}m\frac{2c^3}{kT} \right\} e^{-mc^2/2kT}. \tag{2.59}$$

This expression vanishes if $c = 0$ or $c = \infty$ or $2 = mc^2/kT$, *i.e.*

$$c = \sqrt{\frac{2kT}{m}}. \tag{2.60}$$

If $c < \sqrt{2kT/m}$, then the gradient $d\Pi_c/dc$ in Fig. 2.7 decreases as $c \to \sqrt{2kT/m}$, if $c > \sqrt{2kT/m}$, then the gradient $d\Pi_c/dc$ in Fig. 2.7 increases as $c \to \sqrt{2kT/m}$. Hence Π_c is a maximum when $c = \sqrt{2kT/m}$.

Example 2.7: Probability of a molecule having energy in $(\epsilon, \epsilon + d\epsilon)$[‡‡]
The probability of a molecule having energy between ϵ and $\epsilon + d\epsilon$ is given by

$$\Pi_\epsilon d\epsilon = \frac{2}{\sqrt{\pi}(kT)^{3/2}} e^{-\epsilon/kT} \sqrt{\epsilon} d\epsilon. \tag{2.61}$$

[‡‡]It may be noted that in the following and throughout the entire text the energy of a single element (particle, atom, molecule, oscillator) is denoted by ϵ, whereas the total energy of (say) N elements in a volume V with n_i elements having energy ϵ_i is denoted by $E = \sum_i n_i \epsilon_i$.

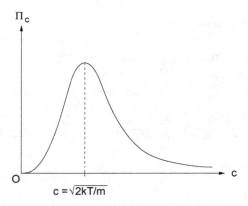

Fig. 2.7 Π_c a maximum at $c = \sqrt{2kT/m}$.

Show that

$$\int_0^\infty \Pi_\epsilon\, d\epsilon = 1. \tag{2.62}$$

Solution: We have:

$$\int_0^\infty \Pi_\epsilon\, d\epsilon = \int_0^\infty \frac{2}{\sqrt{\pi}(kT)^{3/2}} e^{-\epsilon/kT}\sqrt{\epsilon}\, d\epsilon = \frac{2}{\sqrt{\pi}(kT)^{3/2}} \int_0^\infty \sqrt{\epsilon}e^{-\epsilon/kT}\, d\epsilon. \tag{2.63}$$

We set $\epsilon = u^2$, $d\epsilon = 2u\,du$. Then

$$\int_0^\infty \sqrt{\epsilon}e^{-\epsilon/kT}\, d\epsilon = \int_0^\infty ue^{-u^2/kT}2u\,du = 2\int_0^\infty u^2 e^{-u^2/kT}\, du \overset{(2.20)}{=} \frac{2}{4}\sqrt{\pi k^3 T^3}. \tag{2.64}$$

Hence

$$\int_0^\infty \Pi_\epsilon\, d\epsilon = \frac{2\sqrt{\pi}(kT)^{3/2}}{\sqrt{\pi}(kT)^{3/2}2} = 1. \tag{2.65}$$

Example 2.8: Classical theory of thermionic emission

Assume that an electron can only escape from a metal through the potential barrier at the metal surface, located at $x = 0$, if it has a velocity component u in the x-direction such that $mu^2/2 \geq \epsilon_{\text{surface}}$ (actually not all electrons which can escape actually do, and it is necessary to allow for this by multiplying the number n (below) by an efficiency factor). Treating the electrons classically, determine n, the number of electrons which can escape from unit surface area in unit time.

Solution: Using the result (2.42) we have: The number of electrons with x-velocities between u and $u + du$ striking unit area in unit time is

$$N\sqrt{\frac{m}{2\pi kT}}ue^{-mu^2/2kT}du, \tag{2.66}$$

where N is the number of electrons per unit volume in the metal. Therefore the number of electrons which can escape from unit surface area of the metal in unit time is (with $t = mu^2/2kT$, $dt = mudu/kT$, $udu = kTdt/m$)

$$\begin{aligned} n &= N\sqrt{\frac{m}{2\pi kT}} \int_{\sqrt{2\epsilon_{\text{surface}}/m}}^\infty ue^{-mu^2/2kT}\, du \\ &= N\sqrt{\frac{kT}{2\pi m}} \int_{\epsilon_{\text{surface}}/kT}^\infty e^{-t}\, dt = N\sqrt{\frac{kT}{2\pi m}}e^{-\epsilon_{\text{surface}}/kT}. \end{aligned} \tag{2.67}$$

Clearly the correct theory is that which considers the electrons as fermions obeying Fermi–Dirac statistics (see Examples 10.7 to 10.9). The above classical result can therefore be obtained from that of Fermi–Dirac theory in the classical (nondegenerate) limit.[§§]

2.7 Problems without Worked Solutions

Example 2.9: Maxwell averaging of $1/c$

Using the Maxwell distribution for a perfect monatomic gas in thermal equilibrium obtain the average of $1/c$, *i.e.* $\overline{(1/c)}$, where c is the speed of a molecule. How does the result compare with $1/\overline{c}$?

Example 2.10: Probability of one particle picked out from among N

Show that the probability for one particle of a total of N noninteracting particles in a volume V to have a velocity component u_1 in the interval $(u_1, u_1 + du_1)$ is given by the Maxwell distribution law $\Pi_{u_1} du_1$. Start from the following expression

$$
du_1 \int du_2 \dots du_N \, dv_1 \dots dv_N \, dw_1 \dots dw_N \, \delta \left[\frac{1}{2} m(u_2^2 + \dots + w_N^2) - \left(E - \frac{1}{2} mu_1^2 \right) \right]
$$
$$
\sim \theta \left(E - \frac{1}{2} mu_1^2 \right) du_1 \int du \, u^{3N-2} \delta \left[\frac{1}{2} mu^2 - \left(E - \frac{1}{2} mu_1^2 \right) \right], \tag{2.68}
$$

where the second expression results from integration over the angles of the $(3N - 1)$-dimensional unit sphere with u the modulus of $(u_2, \dots, w_N)$. [Hint: Set the argument of the delta function equal to y (say) and integrate and take N and E as large, so that $1 - \alpha/E \simeq \exp(-\alpha/E)$.]

Example 2.11: Standard deviation σ

Consider the probability density Π_u of Eq. (2.50a) and the average $\overline{|u|}$ of Eq. (2.50b). Evaluate correspondingly

$$
\overline{(u - \overline{|u|})^2} = \int_{-\infty}^{\infty} (u - \overline{|u|})^2 \Pi_u \, du = \sigma^2 \tag{2.69}
$$

(the quantity σ being called *standard deviation*). Show that generally:

$$
\left.\begin{aligned}
\overline{(u - \overline{|u|})^{2i}} &= \frac{(2i)!}{2^i i!} \sigma^{2i}, \\
\overline{(u - \overline{|u|})^{2i+1}} &= 0.
\end{aligned}\right\} \quad i = 1, 2, \dots. \tag{2.70}
$$

(Answer: $\sigma^2 = kT/m$).

Example 2.12: Mean energy of effusing molecules

A vapour of molecules of mass m is held at temperature T in some vessel and is allowed to escape through a small hole. What is the mean energy of the molecules emitted from the hole (or through unit area)? (Answer: $kT/2$ multiplied by the particle density).

[§§]The Fermi–Dirac treatment of an electron gas is often described as the degenerate theory.

Chapter 3

The *a priori* Probability

3.1   Introductory Remarks

In this chapter we are concerned with two topics:

1. The *a priori probability*, and

2. insertion of some known physical condition of the system, *e.g.* temperature or energy.

Our considerations in this chapter are almost exclusively classical.

The *a priori* probability as an *a priori*-concept (Latin:from what is before) is defined to be independent of time. We show that the phase-space volume element is such a quantity (a statement known as *Liouville's theorem*). Later, with Heisenberg's uncertainty relation, this quantity is related to a number known in quantum mechanics as *degeneracy*. In the present chapter the classical equivalent is explored and considered in examples by insertion of some known physical condition like the energy. In Chapter 11 the equivalence of the *a priori* probability with degeneracy will become clearer and amounts, in effect, to what is known as the fundamental postulate of the *a priori* probability of an isolated system, a key aspect of statistical mechanics, as discussed in detail in the definitive treatise of R.C. Tolman [78].

3.2   The *a priori* Probability

For an illustration of the concept of *a priori* probability consider the following question in the philosophy of J. Mond [51]. We know there is life on earth. How big was the probability for this event to occur before it occured? It is generally thought that the decisive event took place only once. This would imply (*i.e.* purely on the basis of reasoning, without sensory experience) that

27

the *a priori* probability of this event was practically nought. The probability for a singular event to occur in the universe is practically zero. The classical *a priori* probability* is defined as the ratio of the number of elementary events (*e.g.* the number of times a die is thrown) to the total number of events. In the case of the die each elementary event has the same probability — thus the probability of each outcome of throwing a die is 1/6, as illustrated in Fig. 3.1. Each face of the die appears with equal probability, deduced without a single throw of the die, and therefore independent of time.

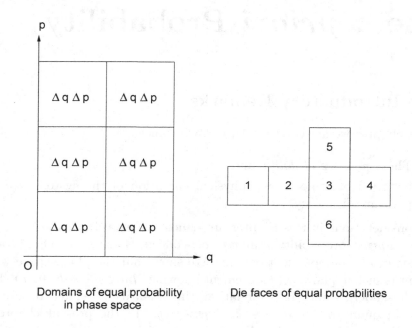

Domains of equal probability   Die faces of equal probabilities
     in phase space

Fig. 3.1 Time-independent *a priori* probabilities.

We consider Hamilton's canonical or phase space coordinates q_i, p_i of a particle i. The dimensions of the phase space are called its degrees of freedom. Consider a particle having a spatial coordinate q and a momentum coordinate p with no other information. We ask: What is the probability of the particle having a spatial coordinate between q and $q + \triangle q$ and a momentum coordinate between p and $p + \triangle p$? We want to show that the *a priori* probability (*i.e.* that with no information about its time dependence and hence no information about its dynamics) is proportional to the phase space volume element, *i.e.*

$$\text{\textit{a priori} probability} \propto \triangle q \triangle p, \tag{3.1}$$

---

*See A. Ben–Naim [4], p.31, M. Toda, R. Kubo and N. Saito [77], p.29.

where $\triangle q$ is the range of the variable q and $\triangle p$ is the range of the variable p.

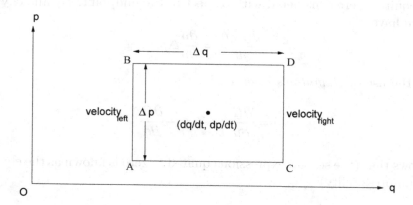

Fig. 3.2 The phase space element $\triangle q \triangle p$ around (q, p).

The answer for an *a priori* probability must be independent of time, since insertion of time would be insertion of known information. For no information on the external conditions we don't know the time. Hence for a consistent choice, time cannot enter, and we have to prove that $\triangle q \triangle p$ is independent of time. This result, *i.e.*

$$\frac{d}{dt} \ln(\triangle q \triangle p) = 0, \qquad (3.2)$$

is known as *Liouville's theorem*. In order to prove this, we observe that

$$\frac{d}{dt} \ln(\triangle q \triangle p) = \frac{d(\triangle q)}{dt} \frac{1}{\triangle q} + \frac{d(\triangle p)}{dt} \frac{1}{\triangle p}. \qquad (3.3)$$

In Fig. 3.2 we show the element $\triangle q \triangle p$ around its center, the phase space point (q, p), which moves through the phase space with velocity $(\dot{q}, \dot{p})$, where $\dot{q} = dq/dt, \dot{p} = dp/dt$. The wall CD in Fig. 3.2 moves with velocity

$$\text{velocity}_{\text{right}} = \dot{q} + \frac{\partial \dot{q}}{\partial q} \frac{\triangle q}{2},$$

and correspondingly the wall AB moves with a velocity

$$\text{velocity}_{\text{left}} = \dot{q} - \frac{\partial \dot{q}}{\partial q} \frac{\triangle q}{2}.$$

Here the rate at which the q-walls move is the difference,

$$\frac{d(\triangle q)}{dt} = \frac{\partial \dot{q}}{\partial q} \triangle q.$$

A velocity is determined by dividing the infinitesimal distance traversed in reaching one point from another (D from B in Fig. 3.2) by the time interval this requires, here considered with respect to the midpoint. By analogy with this we have

$$\frac{d(\triangle p)}{dt} = \frac{\partial \dot{p}}{\partial p}\triangle p.$$

With the use of *Hamilton's equations*

$$\dot{q} = \frac{\partial H}{\partial p}, \qquad \dot{p} = -\frac{\partial H}{\partial q}, \tag{3.4}$$

it follows that (the second expression equated to zero is known as the equation of incompressibility)

$$\frac{d}{dt}\ln(\triangle q \triangle p) = \frac{\partial \dot{q}}{\partial q} + \frac{\partial \dot{p}}{\partial p} = \frac{\partial}{\partial q}\frac{\partial H}{\partial p} - \frac{\partial}{\partial p}\frac{\partial H}{\partial q} = 0. \tag{3.5}$$

This therefore establishes *Liouville's theorem* of the equal probabilities in phase space.[†] What is the analogy with a die? The number on a face of the die, *e.g.* 5, corresponds to the set (q, p) of canonical coordinates in an area h (Planck's constant), generally, of course, in higher dimensions. Just as the number 5 describes one state of the die, so do these coordinates (q, p) in the area h. Just as the faces ("states") of the die with numbers $1, 2, \ldots$ are thrown with the same probability, so similarly the states with different values of (q, p) in areas h occur with the same probability, also called 'coarse grained probability', provided the system is isolated, which means its energy is always the same, the number of different states of the system with the same energy being called its *degeneracy*. Thus we see that the *a priori* probability is related to the concept of degeneracy.

3.3   Examples Illustrating Liouville's Theorem

In the examples we consider now, the Hamiltonian $H(q, p)$ of Hamilton's equations is equal to the energy ϵ. Thus we have

$$\dot{q} = \frac{\partial \epsilon}{\partial p}, \qquad \dot{p} = -\frac{\partial \epsilon}{\partial q}. \tag{3.6}$$

[†]For extensive discussion see J.E. Mayer and M.G. Mayer [45], pp.58-63. See also R.E. Wilde and S. Singh [85], pp.19-20, who make this a postulate, the postulate of equal *a priori* probabilities, which states that for isolated systems at equilibrium, all accessible regions of phase space have equal *a priori* probabilities for equal volumes.

1. *Example*: A particle moving in an electric field with potential energy $V(q)$.

In this case we have

$$\epsilon = \frac{p^2}{2m} + V(q), \tag{3.7}$$

so that

$$\therefore \dot{q} = \frac{\partial \epsilon}{\partial p} = \frac{p}{m} = \frac{m\dot{q}}{m} = \dot{q}, \quad \dot{p} = -\frac{\partial \epsilon}{\partial q} = -\frac{\partial V}{\partial q} = -\text{gradient of } V(q).$$

Here $\dot{p} = \text{force} = \text{electric field in this case.}$

2. *Example*: The photon. In this case we have

$$\epsilon = \hbar c = pc, \quad c = \text{velocity of light,} \tag{3.8}$$

so that

$$\dot{q} = \frac{\partial \epsilon}{\partial p} = c.$$

We see that these Hamiltonian results agree with what one expects. Hence in both cases

$$\frac{d}{dt} \ln(\triangle q \triangle p) = \frac{\partial \dot{q}}{\partial q} + \frac{\partial \dot{p}}{\partial p} = \frac{\partial^2 \epsilon}{\partial q \partial p} - \frac{\partial^2 \epsilon}{\partial p \partial q} = 0.$$

Therefore in both cases $\triangle q \triangle p$ does not change with time. Hence it is sensible to take the *a priori* probability proportional to $\triangle q \triangle p$. We conclude from this, that if this choice is right at some time, it is right at all times.

3.4 Insertion of Physical Conditions

We now consider the insertion of some physical condition of a system, such as its energy. We consider again examples.

1. *Example*: A particle of mass m in a volume V.

In this case we have, as indicated in Fig. 3.3, for a spherical volume,

$$p^2 = p_x^2 + p_y^2 + p_z^2.$$

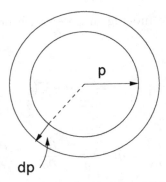

Fig. 3.3 The spherical annulus of thickness dp.

The volume of the spherical annulus shown in Fig. 3.3 is $4\pi p^2 dp$. This is the equivalent element in spherical coordinates. Hence in this case

$$a \ priori \ \text{probability} \ \propto \ V4\pi p^2 dp. \tag{3.9a}$$

If $p \propto c$, the speed of the particle, the *a priori* probability is proportional to $Vc^2 dc$ and (*cf.* Eq. (3.9a)), since

$$\epsilon = \frac{p^2}{2m}, \quad i.e. \quad d\epsilon = \frac{p}{m}dp, \quad \epsilon^{1/2}d\epsilon = \frac{p^2}{m\sqrt{2m}}dp,$$

so that $p^2 dp = \sqrt{2}m^{3/2}\epsilon^{1/2}d\epsilon$, and therefore the *a priori* probability $\triangle V \triangle p_x \triangle p_y \triangle p_z \propto \triangle V p^2 dp$ per particle (if more than one) is proportional to $\triangle V \epsilon^{1/2}d\epsilon$. With division by h^3 (or of that order) the *a priori* probability g becomes the dimensionless quantity

$$g = \frac{1}{h^3}dq_x dq_y dq_z dp_x dp_y dp_z, \tag{3.9b}$$

so that with integration over angles

$$g \to g_p = \frac{1}{h^3}dq_x dq_y dq_z 4\pi p^2 dp = \frac{p^2 dp}{2\pi^2 \hbar^3}dq_x dq_y dq_z,$$

$$g_p \to g_\epsilon = \frac{\sqrt{2}m^{3/2}\epsilon^{1/2}d\epsilon}{2\pi^2 \hbar^3}dq_x dq_y dq_z. \tag{3.9c}$$

We shall see in Chapter 6 that this expression agrees with the accurately derived number of states dN (see after Eq. (6.16)). This agreement shows that indeed the *a priori* probability is to be identified with the number of degenerate states for a specific eigenenergy, *i.e.* it is the number of distinct quantum states with energy ϵ in the interval $d\epsilon$.

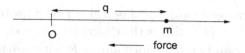

Fig. 3.4 The restoring force of the harmonic oscillator.

2. *Example*: The simple harmonic oscillator.

In this case the particle of mass m is subjected to a restoring force proportional to the spatial coordinate q, as indicated in Fig. 3.4. The energy ϵ consisting of kinetic and potential contributions is given by

$$\epsilon = \frac{p^2}{2m} + 2\pi^2 m\nu^2 q^2, \tag{3.10}$$

where ν is the natural frequency of the oscillator. The *a priori* probability is the probability that the particle has energy between ϵ and $\epsilon + d\epsilon$. We rewrite the above equation in the following form:

$$\frac{p^2}{2m\epsilon} + \frac{q^2}{\epsilon/2\pi^2 m\nu^2} = 1. \tag{3.11}$$

This is the equation of an ellipse of constant energy ϵ, as sketched in Fig. 3.5 for an elliptic annulus of area $\triangle q \triangle p$ bordered by ellipses of energy ϵ and $\epsilon + d\epsilon$. The area of an ellipse with semi-half axes of lengths a and b is known to be given by πab.

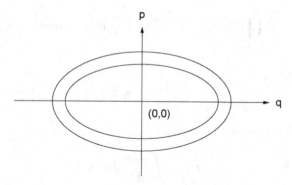

Fig. 3.5 The elliptic annulus in phase space.

Here the area of the ellipse is given by

$$I_1 := \oint dq\,dp = \pi\sqrt{2m\epsilon}\sqrt{\frac{\epsilon}{2\pi^2 m\nu^2}} = \frac{\epsilon}{\nu} \equiv \int \frac{dI_1}{d\epsilon} d\epsilon. \tag{3.12}$$

Thus in the present case the classical *a priori* probability $dqdp$ is proportional to $d\epsilon/\nu$ (*i.e.* this is the difference between the areas of ellipses of energy $\epsilon+d\epsilon$ and ϵ). The normalized density $\rho(q,p)$ of points in phase space is therefore

$$\rho(q,p) = \frac{\nu}{d\epsilon} \quad \text{if} \quad \epsilon < \mathrm{H(q,p)} < \epsilon+d\epsilon, \quad \text{and} \quad 0 \quad \text{otherwise.}$$

Comparing this with the previous case, we observe that here there is no extra energy factor with some power of ϵ. This result will be required and referred to at various points later, *e.g.* in Example 8.3, Eq. (8.24).

3. *Example*: The N-dimensional harmonic oscillator.

As mentioned earlier, the one-dimensional simple harmonic oscillator is not always a good example in statistical physics. Its N-dimensional or N-fold extension is an additional useful example. Thus consider the case of N independent simple harmonic oscillators with total energy

$$E = \sum_i^N \left[\frac{p_i^2}{2m} + 2\pi^2 m \nu^2 q_i^2 \right], \tag{3.13}$$

each oscillator oscillating with the same natural frequency ν. This equation represents an ellipsoid in a $2N$-dimensional hyperspace. The volume of this hyperellipsoid is

$$I_N := \oint \prod_{i=1}^N \{dq_i dp_i\} \propto \left[\sqrt{2mE}\sqrt{E/2\pi^2 m\nu^2} \right]^N \propto E^N. \tag{3.14}$$

We see that

$$\frac{dI_N}{dE} \propto N E^{N-1}. \tag{3.15}$$

Thus in this case we have

$$I_N \propto \int \frac{dI_N}{dE} dE \propto E^N. \tag{3.16}$$

Thus with many oscillators around the *a priori* probability grows in this purely classical consideration with the energy, *i.e.* as E^{N-1}. It is quantum mechanics which prevents this enormous growth with energy provided $N \gg E/h\nu$, as can be inferred from the quantum mechanical treatment (*cf.* Example 12.2, Eq. (12.116)).

We summarize the considerations of the foregoing in the following statements (the phase space coordinates (q, p) always being generalizable to higher dimensions). Although q and p are dynamical (*i.e.* t-dependent) variables, a phase space volume element $\triangle q \triangle p$ is constant in time. The number $\triangle q \triangle p / h$ can be looked at as the number of system points or configurations (or "states") contained in $\triangle q \triangle p$, or correspondingly in the energy interval $\triangle \epsilon \simeq d\epsilon$. In the classical consideration here, these are configurations or states with the same (q, p) (*i.e.* up to the interval $q + \triangle q, p + \triangle p$) and with the same energy ϵ (*i.e.* in the interval $\epsilon, \epsilon + \triangle \epsilon$). In the quantum mechanical treatment the energy is a discrete energy ϵ_i and the number of states with this same energy is what is called the *degeneracy* of the energy level ϵ_i. This degeneracy is denoted by g_i and is therefore classically proportional to the *a priori* probability $\triangle q \triangle p$ or $\triangle \epsilon$, possibly multiplied by some power of ϵ. The examples below illustrate these points. In the literature [73] the lumping together of microscopic, *i.e.* quantum, states is described as 'coarse graining' these to a macroscopic thermal state. Correspondingly 'fine graining' describes the situation of no thermalization, *i.e.* no heat flowing in or out, which leaves the system in a pure state, *i.e.* with no correlation.

3.5 Applications and Examples

Example 3.1: *A priori* weighting for q-independent energies

Show that if the energy of elements is independent of spatial coordinates, the *a priori* weighting g per element is proportional to the accessible volume V.

Solution: We have generally $g_i = \triangle q \triangle p$. Hence $g_i = \triangle q_x \triangle q_y \triangle q_z \triangle p_x \triangle p_y \triangle p_z$. If the energy of the elements is independent of spatial coordinates, it is classically simply $\epsilon = p^2/2m$. Therefore p^2 and hence $\mathbf{p}$ is independent of spatial coordinates, so that g_i is proportional to $\triangle V$ of the volume V, and hence when integrated to V.

Example 3.2: *A priori* weighting in case of angular inclination

Show that if the potential energy of an element depends only upon the angles θ and ϕ of inclination to a given direction, the *a priori* weighting g per element is proportional to $\sin \theta . d\theta d\phi$ in spherical polar coordinates. This case is required later in Example 4.3 which deals with paramagnetism.

Solution: The volume element dV in spherical coordinates is

$$dV = r d\theta . dr . r \sin \theta . d\phi.$$

But the *a priori* weighting g is proportional to the volume, and since the potential energy is stated to be independent of r, the *a priori* weighting g is per element

$$g_i \propto \sin \theta . d\theta . d\phi. \tag{3.17}$$

This result will be used in Example 4.3.

Example 3.3: Gas particles moving in one direction

Show that for a perfect gas confined to movement in one direction of length L, the a priori weighting g per particle is proportional to $L\epsilon^{-1/2}d\epsilon$, where ϵ is the energy of a particle.

Solution: The a priori probability is $g_i = \triangle q \triangle p$. Here $\triangle q = L$ and energy $\epsilon = p^2/2m$ (no change in potential energy). Hence $d\epsilon = pdp/m$, $dp = md\epsilon/p$, and therefore

$$dp = m\frac{d\epsilon}{\sqrt{2m\epsilon}}, \quad \text{and} \quad g \propto L\epsilon^{-1/2}d\epsilon. \tag{3.18}$$

Example 3.4: Perfect gas confined to a plane

For a perfect gas confined to movement on a plane of area A, show that g per particle is proportional to $Ad\epsilon$, ϵ being the energy of a particle.

Solution: Here $g_i = \triangle p \triangle q$ with $dq_x dq_y = A$, and since $\epsilon = p^2/2m$, $p = |\mathbf{p}|$, we have $p = \sqrt{2m\epsilon}$, $dp = md\epsilon/\sqrt{2m\epsilon}$, and $\triangle p_x \triangle p_y \propto pdp \propto d\epsilon$. Thus we obtain

$$g_i \propto Ad\epsilon. \tag{3.19}$$

Example 3.5: Perfect gas confined to volume V

For a perfect gas confined to a volume V, show that g per particle is proportional to $V\epsilon^{1/2}d\epsilon$, ϵ being the energy of a particle.

Solution: This is the case dealt with in the text, first example in Sec. 3.4.

Example 3.6: Phase space of a diatomic molecule

If the rotational energy of a diatomic molecule (i.e. ignoring vibrations along the axis), considered as a rigid rotating dumb-bell as illustrated in Fig. 3.6, and with moment of inertia I is

$$\epsilon = \frac{1}{2I}\left(p_\theta^2 + \frac{p_\phi^2}{\sin^2\theta}\right) \tag{3.20}$$

in spherical polar coordinates, the (p_θ, p_ϕ) curve for constant energy ϵ and angle θ is an ellipse of area

$$\oint dp_\theta dp_\phi = 2\pi I \epsilon \sin\theta. \tag{3.21}$$

By integrating over θ and ϕ, show that the total volume of phase space covered for constant energy ϵ is $8\pi^2 I\epsilon$, and hence $g \propto 8\pi^2 I d\epsilon$.

Fig. 3.6 The diatomic molecule rotating about point O, axis $\perp$ plane of paper.

Solution: We can rewrite the energy of the molecule in the form of the standard equation of a planar ellipse, i.e. as

$$1 = \frac{p_\theta^2}{2I\epsilon} + \frac{p_\phi^2}{2I\epsilon\sin^2\theta}. \tag{3.22}$$

Using the standard formula for the area of an ellipse with semi-major and minor-axes of lengths a and b, i.e. πab, the area A of the ellipse in the present case in the plane of p_θ and p_ϕ of the overall phase space of $\theta, \phi, p_\theta, p_\phi$ is

$$A = \oint dA = \oint dp_\theta dp_\phi d\theta d\phi = \int d\theta d\phi [\pi\sqrt{2I\epsilon}\sqrt{2I\epsilon}\sin\theta] = \int d\theta d\phi [2I\pi\epsilon\sin\theta]. \tag{3.23}$$

Performing the integral over θ, ϕ, we obtain

$$\int_{\phi=0}^{\phi=2\pi} \int_{\theta=0}^{\theta=\pi} 2I\pi\epsilon \sin\theta d\theta d\phi = 2I\pi\epsilon.2\pi.2 = 8\pi^2 I\epsilon. \tag{3.24}$$

It follows that

$$A = 8\pi^2 I\epsilon, \quad dA = 8\pi^2 I d\epsilon, \quad \text{so that} \quad g \propto 8\pi^2 I d\epsilon. \tag{3.25}$$

Related aspects of diatomic molecules are considered in Examples 6.4, 6.8 and 6.9.

Example 3.7: Rhombus boundary as curve of constant energy

If the energy of a certain system per particle is given by

$$\epsilon = \alpha|p| + \beta|q|, \tag{3.26}$$

where α and β are constants, the (q, p) curve for constant ϵ is the boundary of a rhombus of area $\oint dqdp = 2\epsilon^2/\alpha\beta$, as indicated in Fig. 3.7. Hence show that the *a priori* weighting g_i is given by

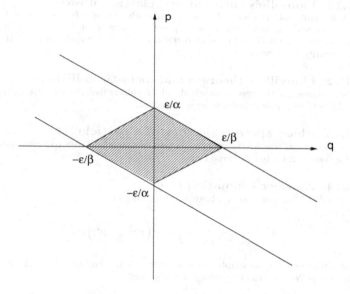

Fig. 3.7 The rhombus given by $\epsilon = \alpha|p| + \beta|q|$.

$g_i \propto 4\epsilon d\epsilon/\alpha\beta$. Note: Here the energy ϵ is linear in q and p, whereas in the example in the text it is quadratic in q and p.

Solution: When $|p| = 0, \epsilon = \beta|q|, |q| = \epsilon/\beta$, i.e. $q = \pm\epsilon/\beta$. Similarly $p = \pm\epsilon/\alpha$ when $|q| = 0$. Any q and any p of the 4 values derived satisfy the above equation. Hence the curve is the boundary of a rhombus. The area A of this rhombus is (twice the area of the triangle in the upper half plane):

$$A = 2\left[\frac{1}{2}2\frac{\epsilon}{\beta}\frac{\epsilon}{\alpha}\right] = \frac{2\epsilon^2}{\alpha\beta}, \quad \text{so that} \quad \oint dqdp = \frac{2\epsilon^2}{\alpha\beta}. \tag{3.27}$$

It follows that g_i is the differential of A, i.e. $g_i = \triangle q\triangle p = (4\epsilon/\alpha\beta)d\epsilon$.

3.6　Problems without Worked Solutions

Example 3.8: System of linear oscillators

Show that for a system of one-dimensional linear harmonic oscillators with coordinate q given by $q = A\cos(\omega t + \theta)$, the phase θ in the range $0 \le \theta \le 2\pi$, the *a priori* probability $P(q)dq$ is proportional to $dq/\sqrt{A^2 - q^2}$, and when normalized to 1 as a proper probability it has $P(q) = 1/\pi\sqrt{A^2 - q^2}$. [Hint: $P(\theta)d\theta = d\theta/2\pi$].

Example 3.9: Throwing a die

How many times must a die be thrown on the average to produce the number 6? [Hint: The probability that in n trials the six appears once and no six $n - 1$ times is $P(n) = (1/6)(5/6)^{n-1}$]. The answer is $\tilde{n} = 6$.

Example 3.10: Time independence of Jacobi determinant

Prove Liouville's theorem by verification of the time independence of the Jacobi determinant J under infinitesimal contact transformations $Q = q + \dot{q}\delta t$, $P = p + \dot{p}\delta t$. [Hint: Start from $\int dQ dP = \int J dq dp$, and prove that $dJ/dt = 0$].

Example 3.11: Liouville's theorem and elastic collisions

Two identical solid balls collide elastically along a smooth, straight furrow. (a) If their velocities before impact are v_1 and v_2, and after impact v_1' and v_2', show that $dv_1 dv_2 = dv_1' dv_2'$. (b) Verify Liouville's theorem for the elastic collision of two solid balls of unequal masses. [Hint: Start from momentum and energy conservations].

Example 3.12: Liouville's theorem and inelastic collisions

Verify Liouville's theorem in the case of two balls which collide inelastically (*i.e.* after the collision the dimensionality of the phase space has been reduced).

Example 3.13: Phase space trajectory of a particle

What is the (x, p_x) phase space trajectory of a particle which is free to move along the x-axis between 0 and a, and is reflected at these ends?

Example 3.14: Another evaluation of g_ϵ

Show that the result (3.9c) can also be obtained by evaluation of

$$g_\epsilon = \frac{1}{(2\pi\hbar)^3} \int_V d\mathbf{q} \int d\mathbf{p}\delta\left(\epsilon - \frac{1}{2m}\mathbf{p}^2\right)d\epsilon, \qquad (3.28)$$

the delta function representing the number of states per unit interval of ϵ. What is the ϵ-dependence of g_ϵ in the case of an N-dimensional configuration space?

Example 3.15: Ratio of statistical weights

Consider a single α-particle trapped within a sphere of radius a representing a nucleus. In the Coulomb region $r > a$ the potential barrier falls off as $V = +2Ze^2/r$; inside the sphere $V = 0$. The α-particle penetrates the barrier from $r = a$ to $r = b$, $b > a$, with energy E. What is the ratio of the statistical weight at $r = b$ to that at $r = a$? (Answer: b^2/a^2)

Chapter 4

Classical Statistics (Maxwell–Boltzmann)

4.1 Introductory Remarks

In the following we consider the *method of the most probable distribution* which — as Schrödinger remarks[*] recommends itself by its great simplicity — which, however, to make it entirely satisfactory, would require a rigorous proof that for $N \to \infty$ the deviations from the most probable distribution can be rigorously neglected. A different and rigorous method has therefore also been developed; this is the *method of mean values* to be discussed in Chapter 12. The differences between the two methods have been emphasized particularly by Schrödinger [65] and Dingle [20].

4.2 The Number of Arrangements of Elements in Maxwell–Boltzmann Statistics

Classical or *Maxwell–Boltzmann statistics* is based on three assumptions. These are:

1. All *elements* in a system are *distinguishable* (elements of the system being individual members, *i.e.* particles or the like, as photons).

2. All "*states*" (*i.e.* all energy "levels") are also distinguishable, even if degenerate, *degeneracy* meaning the number of different states with the same energy. Of course, in classical statistics we do not have a countable number of discrete energy levels ϵ_i as in quantum mechanics.

---

[*]E. Schrödinger [65], p.22.

Instead we have *e.g.* as the energy of a mass m in a box of volume V or in a potential $V(q)$ the energy $\epsilon_i = p^2/2m + V(q)$. Nonetheless it is convenient to think of the levels as discrete in this classical context in the sense that all energies ϵ in an interval $(\epsilon, \epsilon + d\epsilon)$ are lumped together and considered as the one energy level ϵ.

3. Any state can accommodate *any number of elements*.

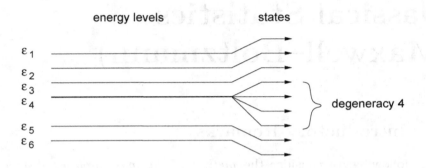

Fig. 4.1 An energy spectrum with one level of 4-fold degeneracy.

We consider energy levels ϵ_i with $i = 1, 2, \ldots$.[†] We assume level i is degenerate, *i.e.* has several different states (solutions of the Schrödinger equation, as indicated in Fig. 4.1) with the same energy ϵ_i, and has n_i elements (*i.e.* the system consisting of many identical and independent particles or atoms has n_i of these with the same single particle energy ϵ_i, which is why n_i is also called an "*occupation number*", since these particles occupy the level ϵ_i). We denote by g_i this number called the *degeneracy of energy level i*. This *degeneracy* is here in classical physics (proportional to) the *classical a priori probability*, *i.e.* (with proportionality constant $1/h$ here put equal to 1)

$$g_i := \text{degeneracy} \equiv \triangle q \triangle p. \qquad (4.1)$$

We suppose we have in a particular distribution n_i elements distributed amongst the g_i states of the energy level ϵ_i. We assume that all states and elements are *distinguishable*. Then, considering the *number of arrangements* of the n_i distinguishable elements amongst the g_i distinguishable states of

---

[†]Note that here we begin counting from 1. We want to reserve 0 as index of the energy called *Fermi energy* below. Later we shall use ϵ_0 also as ground state energy, *cf.* Chapter 12. The relevant meaning of ϵ_0 is to be deduced from the context.

the energy level ϵ_i, we have:

> The first element can go in g_i places (representing states),
>
> the second element can go in g_i places,
>
> $\ldots\ldots\ldots\ldots\ldots\ldots\ldots\ldots$,

because we can put any number of elements into the same place. It follows that the number of possible arrangements is $g_i \times g_i \times \cdots g_i = g_i^{n_i}$. Now suppose we have a *total number of N* elements, $N = \sum_i n_i$. The number of ways of dividing N elements into packets of

> n_1 on energy level ϵ_1,
>
> n_2 on energy level ϵ_2,
>
> $\ldots\ldots\ldots\ldots\ldots\ldots\ldots$,
>
> n_i on energy level ϵ_i,
>
> $\ldots\ldots\ldots\ldots\ldots\ldots\ldots$,

is equal to:

$$= \frac{N!}{n_1! n_2! n_3! \cdots n_i! \cdots} = N! \prod_i \frac{1}{n_i!}.$$

Taking into account the degeneracy g of every level, it follows therefore that:

> The total number of arrangements giving
>
> n_1 elements on ϵ_1,
>
> n_2 elements on ϵ_2,
>
> n_3 elements on ϵ_3,
>
> $\ldots\ldots\ldots\ldots\ldots\ldots$

with $N = \sum_i n_i$ is the *Maxwell–Boltzmann number of arrangements of particles*

$$W_{MB} = N! \prod_i \frac{g_i^{n_i}}{n_i!}. \tag{4.2}$$

This number of distinguishable arrangements corresponds in quantum statistics to the number of accessible states, as will become clear later.

4.3   Method of Maximum Probability

The *method of maximum probability* consists of finding particular values of $n_1, n_2, n_3, \ldots, n_i, \ldots$, which lead to the maximum number of arrangements, since the resultant system will turn up most frequently. Therefore we consider $\ln W_{MB}$ and *maximize* it. It is a basic assumption of the method that the total number of elements N is huge. Thus a rigorous proof of the validity of the method would require a demonstration that deviations from the most probable distribution can be rigorously neglected.[‡] In any case, one would like to have a measure of how sharp the maximum is. This is investigated *e.g.* in the book of Reif [59], Sec. 3.7.

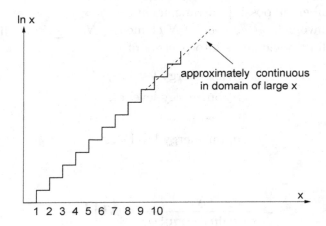

Fig. 4.2 $\ln x$, where x is regarded as a continuous variable for large x.

We have:

$$\ln W_{MB} = \ln N! + \sum_i [n_i \ln g_i - \ln n_i!]. \tag{4.3}$$

We next search for an approximation to $\ln n!$. The result is known as *Stirling's theorem* or Stirling's approximation. We have (*cf.* Fig. 4.2)

$$\ln n! = \ln[1 \times 2 \times 3 \times 4 \ldots \times n] = \sum_{x=1}^{n} \ln x = \sum_{x=1}^{n} 1 \times \ln x$$

$$= \text{area under the staircase in Fig. 4.2.} \tag{4.4}$$

Thus to a good approximation:

$$\ln n! \simeq \int_{x=1}^{x=n} \ln x \, dx$$

$$= [x(\ln x - 1)]_1^n$$

$$\simeq n(\ln n - 1), \quad \text{where } n \text{ is large.} \tag{4.5}$$

[‡]E. Schrödinger [65], p.22.

Applying this result to W_{MB}, we have therefore:

$$\ln W_{MB} \simeq \ln N! + \sum_i [n_i \ln g_i - n_i \ln n_i + n_i]. \tag{4.6}$$

In order to determine the maximum value (check the second variation) we differentiate this expression and equate the result to zero. Then

$$0 = d(\ln W_{MB}) = \sum_i dn_i[\ln g_i - \ln n_i - 1 + 1]$$

$$= -\sum_i dn_i \ln \left(\frac{n_i}{g_i}\right). \tag{4.7}$$

We assume the energy E of the system is known. Then we have the additional conditions:

$$\text{const.} = E = \sum_i n_i \epsilon_i, \quad dE = 0 = \sum_i dn_i \epsilon_i. \tag{4.8}$$

In the rest of the text the maximization will mostly be forgotten — but see Example 5.6.

4.3.1 The case of nonconserved elements

As a first case we consider that of nonconserved elements. This is the case of *e.g. photons* and *phonons*. *Nonconservation* here means there is no restriction on the number.[§] Applying the method of *Lagrangian multipliers* to Eqs. (4.7) and (4.8), we obtain

$$\ln \left(\frac{n_i}{g_i}\right) = -\mu \epsilon_i, \tag{4.9}$$

where μ is the Lagrangian multiplier and is a constant for the system. It follows that we have:

$$n_i = g_i e^{-\mu \epsilon_i}. \tag{4.10}$$

We repeat: This is the *Maxwell–Boltzmann distribution* if the elements are *not conserved*. Note the "equal-to" sign, not proportional.

4.3.2 The case of conserved elements

In the case of conserved elements, *i.e.* if their number is restricted to (say) N, we have the additional condition (*cf.* Eq. (2.31))

$$N = \sum_i n_i, \quad 0 = \sum_i dn_i. \tag{4.11}$$

[§]Thus *e.g.* the atoms of the walls of the container absorb and emit photons in many different ways so that the number of photons in the container is not fixed. The same applies to phonons.

Again using the method of Lagrangian multipliers, *cf.* Eqs. (2.31), (2.34), we obtain from Eqs. (4.7), (4.8) and (4.11) the relation:

$$\ln\left(\frac{n_i}{g_i}\right) = -\mu\epsilon_i + \text{const.}, \qquad \text{const.} \equiv -A' \equiv \mu\epsilon_0. \tag{4.12}$$

Hence in this case we have (absorbing the additive constant $-A'$ in Eq. (4.12) in the multiplicative constant in the following — this is different later in quantum statistics of conserved elements, *cf.* Eqs. (7.9) to (7.11)):

$$n_i \propto g_i e^{-\mu\epsilon_i}, \qquad n_i = \frac{g_i}{e^{\mu(\epsilon_i-\epsilon_0)}}. \tag{4.13}$$

It follows that the total number N is:

$$N = \sum_i n_i = \sum_i g_i e^{-\mu(\epsilon_i-\epsilon_0)},$$

and therefore the ratio of the last two relations implies (note that here ϵ_0 cancels out)[¶]

$$n_i = N\frac{g_i e^{-\mu(\epsilon_i-\epsilon_0)}}{\sum_i g_i e^{-\mu(\epsilon_i-\epsilon_0)}} = N\frac{g_i e^{-\mu\epsilon_i}}{\sum_i g_i e^{-\mu\epsilon_i}}. \tag{4.14}$$

This result is the occupation number n_i of the *most probable distribution* of particles amongst the single-particle energy levels ϵ_i. We shall later obtain the same expression for the mean value with the rigorously correct method of mean values (see Eq. (12.19)).

4.3.3   The meaning of μ

In order to extract the meaning of the constant μ, we prove first that two systems in thermal contact have the same μ. Consider the

first system with :

N elements, energy levels ϵ_i, with g_i states and n_i elements on ϵ_i,

and the

second system with :

$\tilde{N}$ elements, energy levels $\tilde{\epsilon}_j$, with $\tilde{g}_j$ states, and $\tilde{n}_j$ elements on $\tilde{\epsilon}_j$.

[¶]We mention that — different from us here — in the book of D.C. Mattis and R.H. Swendsen [44], p.15, the Boltzmann factors $\exp(-\mu\epsilon_i)$ are said to be *a priori* probabilities.

The number of arrangements of the combined system is the product $W_{MB} = W_{MB}^{(1)} W_{MB}^{(2)}$. Hence

$$W_{MB} = N! \tilde{N}! \prod_i \frac{g_i^{n_i}}{n_i!} \prod_j \frac{\tilde{g}_j^{\tilde{n}_j}}{\tilde{n}_j!}, \qquad (4.15a)$$

and thus for the determination of the maximum number:

$$0 = d\ln W_{MB} = d\ln W_{MB}^{(1)} + d\ln W_{MB}^{(2)}$$

$$\stackrel{(4.7)}{=} -\left[\sum_i dn_i \ln\left(\frac{n_i}{g_i}\right) + \sum_j d\tilde{n}_j \ln\left(\frac{\tilde{n}_j}{\tilde{g}_j}\right) \right]. \qquad (4.15b)$$

We know only the energy E of the systems combined since they are in thermal contact, so that

$$E = \sum_i n_i \epsilon_i + \sum_j \tilde{n}_j \tilde{\epsilon}_j, \quad 0 = dE = \sum_i dn_i \epsilon_i + \sum_j d\tilde{n}_j \tilde{\epsilon}_j. \qquad (4.16)$$

The Lagrangian multiplier of Eqs. (4.15b) and (4.16) is μ, *i.e.* equal to the same as before in view of the independence of $dn_i, d\tilde{n}_j$. Therefore for *non-conserved systems* like photons (we do not have a fixed number of these) we have

$$\ln\left(\frac{n_i}{g_i}\right) = -\mu\epsilon_i \quad \text{and} \quad \ln\left(\frac{\tilde{n}_j}{\tilde{g}_j}\right) = -\mu\tilde{\epsilon}_j. \qquad (4.17)$$

If the systems are *thermally isolated*, and if the energy of each is known, then

$$0 = \sum_i dn_i \epsilon_i \quad \text{and} \quad 0 = \sum_j d\tilde{n}_j \tilde{\epsilon}_j. \qquad (4.18)$$

We then obtain a different Lagrangian multiplier μ for each system.

4.3.4 Identification of μ with 1/kT

The system of interest is assumed to be *in thermal contact with the system consisting of a perfect gas* — *i.e.* they both have the same μ's and are in a state of equilibrium. We now determine μ for a perfect gas. We have

$$\text{probability} \propto g_i e^{-\mu\epsilon_i}.$$

We write as earlier g as $V dp_x dp_y dp_z$ with

$$\epsilon = \frac{1}{2m}(p_x^2 + p_y^2 + p_z^2).$$

This probability is composed of three separate factors, one for each direction, *e.g.*

$$\text{probability}_{(p_x, p_x + dp_x)} \propto e^{-\mu p_x^2/2m}.$$

Maxwell assumed that the three orthogonal directions behaved independently; now shown to be true. Consider the pressure P on the wall perpendicular to the x-direction. We have (*cf.* Eq. (2.39))

$$P = \frac{\text{force}}{\text{area}} = \text{force per unit area} = N \frac{\int_0^\infty (p_x/m) 2 p_x e^{-\mu p_x^2/2m} dp_x}{\int_{-\infty}^\infty e^{-\mu p_x^2/2m} dp_x}, \quad (4.19a)$$

where N is the number of molecules per unit volume, and $2p_x$ is the momentum change per particle at the wall. Note the integral in the numerator starts from 0 because the particles must be moving towards the wall if they are to contribute to the pressure P. The factor p_x/m is the velocity perpendicular to the wall — here required to give the number of molecules hitting unit area in unit time. The integral in the denominator is the normalization factor. Evaluating the integrals with results of Example 2.1, we obtain:

$$P = N \frac{2}{m} \frac{\int_0^\infty p_x^2 e^{-\mu p_x^2/2m} dp_x}{\int_{-\infty}^\infty e^{-\mu p_x^2/2m} dp_x} \overset{(2.20)}{=} N \frac{2}{m} \frac{[(1/4)\sqrt{8\pi m^3/\mu^3}]_0^\infty}{2[(1/2)\sqrt{\pi 2m/\mu}]_0^\infty} = \frac{N}{\mu}. \quad (4.19b)$$

But for the *perfect gas*: $P = NkT$.[||] Therefore

$$\mu = \frac{1}{kT}. \quad (4.20)$$

Recall here again, the result

$$n_i = N \frac{g_i e^{-\epsilon_i/kT}}{\sum_i g_i e^{-\epsilon_i/kT}} \quad (4.21)$$

is the *Maxwell–Boltzmann distribution law* for *conserved elements* (N elements). The factor $\exp(-\epsilon_i/kT)$ represents a single particle energy weighting factor. The sum

$$\sum_i g_i e^{-\epsilon_i/kT}$$

is described as an *internal partition function*[**] because *e.g.* the additive translational and rotational parts of ϵ_i permit a product partition of the exponentials, which is a very common feature of such terms that we shall encounter throughout the text.

[||] One learns the formula as $PV = NkT$. In the present context $V = 1$, *i.e.* N is the number of molecules per unit volume, *i.e.* Avogadro's number, and $R = Nk$ is the universal gas constant, where $k = 1.38 \times 10^{-23} J/K$ is the Boltzmann constant.

[**] For instance by O.K. Rice [60], p.43.

4.3.5 Distribution of particles in the atmosphere

The probability of a particle of mass m and energy ϵ_i to be at a height h in the atmosphere is now given by the expression (4.13),

$$\text{probability } P(h) \propto g_i e^{-\epsilon_i/kT}, \quad \text{where} \quad \epsilon_i = \frac{\mathbf{p}^2}{2m} + mgh, \qquad (4.22)$$

where g is the acceleration due to gravity. The momentum distribution is independent of the height h. Therefore

$$g_i \propto V \triangle p,$$

where V is the volume considered. Hence if $N(h)$ is the number of particles per unit volume at height h, the probability of having a particle at height h, $P(h)$, is

$$\propto e^{-mgh/kT},$$

because the momentum variation $\triangle p$ is independent of the height h. If the temperature T is constant with h,

$$P(h) \propto N(h)e^{-mgh/kT}, \qquad P(h) = P(0)e^{-mgh/kT}. \qquad (4.23)$$

This is an example illustrating the application of the Maxwell–Boltzmann distribution law. The formula (4.23) is known as the *barometric formula* for a height h.

4.3.6 Law of equipartition of energy

Energy contains squared terms of the form, *e.g.*

$$\epsilon_{p_x} = \frac{1}{2m}p_x^2 \quad \text{(linear translation)},$$

$$\epsilon_{p_\phi} = \frac{1}{2I}p_\phi^2 \quad \text{(rotation, I} = \text{moment of inertia)},$$

$$\epsilon_{q_x} = 2\pi^2 m\nu^2 q_x^2 \quad \text{(energy of vibration of freqency } \nu). \qquad (4.24)$$

Here for p_x, p_ϕ, q_x the domains are $-\infty$ to ∞. The corresponding probability distributions are

$$\text{probability} \propto g_i e^{-\epsilon_i/kT},$$

where

$$g_i \propto dq_x dp_x dq_y dp_y \quad \text{or} \quad dq_\phi dp_\phi,$$

and the energy ϵ is the sum of squared terms. Thus the corresponding probabilities are respectively proportional to

$$dp_x e^{-p_x^2/2mkT}, \quad dp_\phi e^{-p_\phi^2/2IkT}, \quad dq_x e^{-2\pi^2 m\nu^2 q_x^2/kT}$$

(relevant factors all multiplied together).

The average or mean energy $\overline{E}$ of a system is now (in the present context) defined by the relation

$$\overline{E} = \frac{\sum_i \epsilon_i g_i e^{-\epsilon_i/kT}}{\sum_i g_i e^{-\epsilon_i/kT}}. \tag{4.25}$$

Other means are formed correspondingly. Consider, for example, a case

$$\text{probability} \propto dp e^{-ap^2/kT} \tag{4.26a}$$

corresponding to contribution ap^2 to energy ϵ. The *mean energy contribution* is then given by (the denominator being the normalization factor)

$$\begin{aligned}
\text{mean contribution} &= \frac{\int_{-\infty}^{\infty}(ap^2)e^{-ap^2/kT}\,dp}{\int_{-\infty}^{\infty}e^{-ap^2/kT}\,dp} \\
&= \frac{2a(1/4)\sqrt{T^3\pi k^3/a^3}}{2(1/2)\sqrt{\pi Tk/a}} \\
&= \frac{1}{2}kT.
\end{aligned} \tag{4.26b}$$

Thus every squared term of the motion in the energy of a particle gives a mean contribution $kT/2$ to the total energy of the system. These considerations depend on the following assumptions:

1. All variables are assumed to be continuous (this breaks down in quantum theory).

2. The limits of the p's and q's are $\pm\infty$, *i.e.* the limits of ϵ are $+\infty$ at both ends.

3. The distribution $e^{-\epsilon/kT}$ is true in the classical theory only.

4.4 Applications

As applications we consider the specific heat or heat capacity for a monatomic gas and a solid. (Diatomic molecules are treated in Examples 6.8 and 6.9).

4.4.1 The monatomic gas

In the case of a free monatomic gas we have three terms of linear translation (a phase space of 3 dimensions):

$$\epsilon = \frac{1}{2m}(p_x^2 + p_y^2 + p_z^2). \tag{4.27a}$$

It follows that the mean energy contribution per particle is

$$\epsilon = 3 \times \frac{1}{2}kT = \frac{3}{2}kT \text{ per particle.} \tag{4.27b}$$

The *specific heat* or *capacity of heat* of N particles[tt] is therefore with $E = N\epsilon$

$$C_V = \left(\frac{\partial E}{\partial T}\right)_V = \frac{\partial N\epsilon}{\partial T} = \frac{3}{2}Nk, \tag{4.28}$$

where N is the number of particles. For 1 gm molecule, the specific heat at constant volume V is[tt] $C_V = 3R/2, R = Nk$.

4.4.2 A solid

In the case of a solid we have 6 energy terms: 3 terms of linear translation and 3 terms of vibration. It follows that the mean energy per particle is

$$\epsilon = 6 \times \frac{1}{2}kT = 3kT,$$

and therefore the specific heat is

$$C_V = 3Nk = 3R \text{ per gm molecule.} \tag{4.29}$$

This result is known as the *law of Dulong and Petit*.

If we reduce the general quadratic form of ϵ to the sum of perfect squares, the number of perfect squares in E is $6N$, where N is the number of atoms. The quantum statistical treatment of the specific heat of solids is given in Chapter 9.

4.5 Applications and Examples

Example 4.1: Euler–Maclaurin summation formula
Show that the *Euler–Maclaurin summation formula*[§§]

$$\sum_{n=a}^{b} f(n) = \int_a^b f(n)dn + \frac{1}{2}[f(a) + f(b)] + \frac{1}{12}[f'(b) - f'(a)] + \cdots, \tag{4.30}$$

[tt]If a macroscopic system is given the amount dQ of heat at constant macroscopic parameter y (e.g. V or P) and with temperature change dT, one defines as the capacity of heat of the system the quantity $C_y = (dQ/dT)_y$. With $dQ = TdS$, S the entropy (see Chapter 5) we also have $C_y = T(\partial S/\partial T)_y$. The specific heat per unit mass is $c_y = C_y/Nm$, where m is the mass of one molecule of the gas. From the second law of thermodynamics, $dQ = dE + PdV = TdS$, we obtain $C_V = (\partial E/\partial T)_V = T(\partial S/\partial T)_V$.

[tt]Thus $C_V T$ represents roughly the heat content of a macroscopic system for moderate temperatures T.

[§§]This formula is discussed in J.E. Mayer and M.G. Mayer [45], pp.152-154. For the derivation see pp.431-432. It is a formula used for numerical integration. With more terms, and $d^{(i)} := [f^{(i)}(b) - f^{(i)}(a)]$, it is $\int_a^b f(n)dn = \sum_{n=a}^b f(n) - \frac{1}{2}[f(a) + f(b)] - \frac{1}{12}d^{(1)} + \frac{1}{720}d^{(3)} - \frac{1}{30240}d^{(5)} + \frac{1}{1209600}d^{(7)} - \frac{1}{1197504000}d^{(9)} + \cdots$.

when applied to the identity (compare with Eq. (4.4))

$$\ln n! = \sum_{n=1}^{n} \ln n, \tag{4.31}$$

gives

$$\ln n! \simeq \left(n + \frac{1}{2}\right) \ln n - n + A. \tag{4.32}$$

Show that the above terms of the formula lead to the value $A \approx 11/12$.

Solution: We have:

$$
\begin{aligned}
\sum_{n=1}^{n} \ln n &= \int_{1}^{n} \ln n\, dn + \frac{1}{2}[\ln 1 + \ln n] + \frac{1}{12}\left[\frac{d}{dn}\ln n\right]_{1}^{n} + \cdots \\
&\simeq [n(\ln n - 1)]_{1}^{n} + \frac{1}{2}\ln n + \frac{1}{12}\left[\frac{1}{n} - 1\right] \\
&\simeq n\ln n - n + 1 + \frac{1}{2}\ln n + \frac{1}{12}\frac{1}{n} - \frac{1}{12} \\
&\simeq \left(n + \frac{1}{2}\right)\ln n - n + A,
\end{aligned}
\tag{4.33}
$$

for n large so that $1/12n \to 0$, where $A = 1 - 1/12 = 11/12$. The coefficients in Eq. (4.30) are identical with those of the expansion

$$\sum_{j=0}^{\infty} e^{-aj} = \frac{1}{1 - e^{-a}} = \frac{1}{a} + \frac{1}{2} + \frac{1}{12}a - \frac{1}{720}a^3 + \cdots. \tag{4.34}$$

Example 4.2: Comparison of Euler–Maclaurin and Stirling's formulas

Compare the approximate result of Example 4.1 with that of Stirling's formula (*cf.* Example 12.8)

$$n! \simeq \sqrt{2\pi}\, n^{n+1/2} e^{-n}. \tag{4.35}$$

Solution: Taking the logarithm of Stirling's formula we obtain

$$\ln n! = \ln \sqrt{2\pi} + n\ln n - n + \frac{1}{2}\ln n. \tag{4.36}$$

Therefore

$$\ln n! \simeq \left(n + \frac{1}{2}\right)\ln n - n + \ln \sqrt{2\pi}. \tag{4.37}$$

But $\sqrt{2\pi} \simeq \sqrt{6.28} \simeq 2.506$, so that $\ln \sqrt{2\pi} \simeq 0.9187$, which is to be compared with $11/12 \simeq 0.9166$ in Example 4.1. Hence the results are almost identical (*i.e.* true to two decimal places). Note that if one keeps only the integral in Eq. (4.30), one has $\ln n! \simeq n(\ln n - 1) + 1$. Even that is in many cases not a bad approximation!

Example 4.3: Langevin formula for paramagnetism

Demonstrate from the Maxwell–Boltzmann distribution law that the probability of a magnetic moment M setting at an angle θ to the direction of a magnetic field H is proportional to

$$e^{MH \cos \theta / kT} \sin \theta\, d\theta. \tag{4.38}$$

Hence deduce that the mean magnetic moment per element (*i.e.* for N elements the result would contain a factor N) is

$$\overline{M} = M\left[\coth\left(\frac{MH}{kT}\right) - \frac{1}{MH/kT}\right].\tag{4.39}$$

Show that this reduces for small H to

$$\overline{M} \simeq \frac{1}{3}\frac{M^2 H}{kT}.\tag{4.40}$$

This result is known as the *Langevin formula for paramagnetism*.

Solution: The field $\mathbf{H}$ exerts a couple or torque $\mathbf{M} \times \mathbf{H}$ on $\mathbf{M}$. For alignment of $\mathbf{M}$ with $\mathbf{H}$ at angle $\theta = 0$ the energy is $\epsilon = -MH$. We have therefore

$$n_i \propto g_i e^{-\epsilon_i/kT} \quad \text{with} \quad \epsilon_i = -\mathbf{M}\cdot\mathbf{H} = -MH\cos\theta.\tag{4.41}$$

From Example 3.2 we know that $g_i \propto \sin\theta d\theta d\phi$. It follows that

$$n_i \propto e^{MH\cos\theta/kT}\sin\theta d\theta,\tag{4.42}$$

since ϕ is constant (symmetry about the $\mathbf{H}$-axis). We have therefore as average (like that of Eq. (4.25)):

$$\overline{M} := M\frac{\int_0^\pi \cos\theta\sin\theta e^{MH\cos\theta/kT}d\theta}{\int_0^\pi e^{MH\cos\theta/kT}\sin\theta d\theta}.\tag{4.43}$$

Consider first the integral in the denominator:

$$\int_0^\pi e^{MH\cos\theta/kT}\sin\theta d\theta = -\left[\frac{kT}{MH}e^{MH\cos\theta/kT}\right]_0^\pi = \frac{kT}{MH}\left[e^{MH/kT} - e^{-MH/kT}\right].\tag{4.44}$$

We evaluate the integral in the numerator by partial integration:

$$\int_0^\pi \cos\theta\sin\theta e^{MH\cos\theta/kT}d\theta$$

$$= \left[\cos\theta\int^\theta \sin\theta e^{MH\cos\theta/kT}d\theta\right]_0^\pi - \int_0^\pi(-\sin\theta)\left[\int^\theta \sin\theta e^{MH\cos\theta/kT}d\theta\right]d\theta$$

$$= \left[\cos\theta\int^\theta \sin\theta e^{MH\cos\theta/kT}d\theta\right]_0^\pi + \int_0^\pi(-\sin\theta)\left\{\frac{kT}{MH}e^{MH\cos\theta/kT}\right\}d\theta$$

$$= -\left[\frac{kT\cos\theta}{MH}e^{MH\cos\theta/kT}\right]_0^\pi - \frac{kT}{MH}\left[-\frac{kT}{MH}e^{MH\cos\theta/kT}\right]_0^\pi$$

$$= \frac{kT}{MH}e^{-MH/kT} + \frac{kT}{MH}e^{MH/kT} + \left(\frac{kT}{MH}\right)^2 e^{-MH/kT} - \left(\frac{kT}{MH}\right)^2 e^{MH/kT}.\tag{4.45}$$

It follows that

$$\overline{M} = M\frac{e^{-MH/kT} + e^{MH/kT} - \frac{kT}{MH}e^{MH/kT} + \frac{kT}{MH}e^{-MH/kT}}{e^{MH/kT} - e^{-MH/kT}}$$

$$= M\left[\frac{e^{MH/kT} + e^{-MH/kT}}{e^{MH/kT} - e^{-MH/kT}} - \frac{kT}{MH}\right]$$

$$= M\left[\coth\frac{MH}{kT} - \frac{kT}{HM}\right].\tag{4.46}$$

Finally we expand $\coth x = \cosh x / \sinh x$ in rising powers of the argument $x = MH/kT$, *i.e.* for small H. Then the average magnetic moment per element is

$$
\begin{aligned}
\overline{M} &= M \left[\frac{1 + (MH/kT)^2/2! + (MH/kT)^4/4! + \cdots}{(MH/kT) + (MH/kT)^3/3! + \cdots} - \frac{kT}{MH} \right] \\
&\simeq M \left[\frac{1 + M^2H^2/2k^2T^2 - M^2H^2/6k^2T^2}{MH/kT} - \frac{kT}{MH} \right] \simeq \frac{M^2 H}{3kT}.
\end{aligned}
\tag{4.47}
$$

This result — achieved with Maxwell–Boltzmann statistics — will later in Example 5.3 be rederived in the context of thermodynamics, and again later in Example 10.6 it will be compared with that derived with Fermi–Dirac statistics, which is therefore called the *Pauli spin paramagnetism* of a metal.

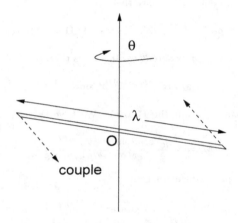

Fig. 4.3 The restoring couple acting on the needle.

Example 4.4: Root-mean-square deflection of a suspended needle

A uniform needle of mass M and length λ is suspended at its centre by a thin torsion thread which exerts a restoring couple $a\theta$ or twist through an angle θ. Show that the root-mean-square angular deflection of the needle and its root-mean-square angular velocity are respectively given by

$$
\sqrt{\overline{\theta^2}} = \sqrt{\frac{kT}{a}} \quad \text{and} \quad \sqrt{\overline{\dot\theta^2}} = \frac{2}{\lambda}\sqrt{\frac{3kT}{M}}.
\tag{4.48}
$$

Solution: The moment of inertia I of a rod of length λ rotated around an axis through its centre is $I = M(\lambda/2)^2/3$, as indicated in Figs. 4.3 and 4.4. The relevant Lagrangian $\mathcal{L}$ is

$$
\mathcal{L}(\theta, \dot\theta) = \frac{1}{2} I \dot\theta^2 - \frac{1}{2} a \theta^2,
\tag{4.49}
$$

with equation of motion

$$
\frac{d}{dt}\frac{\partial \mathcal{L}}{\partial \dot\theta} - \frac{\partial \mathcal{L}}{\partial \theta} = 0, \quad I\ddot\theta + a\theta = 0.
\tag{4.50}
$$

The energy of the needle is

$$
\epsilon = \frac{1}{2} I \dot\theta^2 + \frac{1}{2} a \theta^2 = \frac{L^2}{2I} + \frac{1}{2} a \theta^2, \quad \text{where} \quad L = I\dot\theta.
\tag{4.51}
$$

The Maxwell–Boltzmann energy weighting factor is therefore

$$e^{-\epsilon/kT} = e^{-(I\dot{\theta}^2/2 + a\theta^2/2)/kT}.$$

Thus angular deflection θ and angular momentum L separate. The mean-square deflection is therefore given by

$$\overline{\theta^2} = \frac{\int_0^\infty \theta^2 e^{-a\theta^2/2kT}\,d\theta}{\int_0^\infty e^{-a\theta^2/2kT}\,d\theta} = \frac{\int_0^\infty \theta^2 e^{-\alpha\theta^2}\,d\theta}{\int_0^\infty e^{-\alpha\theta^2}\,d\theta}, \qquad \text{where} \qquad \alpha = \frac{a}{2kT},$$

$$\overset{(2.20)}{=} \frac{1}{4}\sqrt{\frac{\pi}{\alpha^3}}\frac{2}{1}\sqrt{\frac{\alpha}{\pi}} = \frac{1}{2\alpha} = \frac{kT}{a}. \tag{4.52}$$

Thus finally

$$\sqrt{\overline{\theta^2}} = \sqrt{\frac{kT}{a}}. \tag{4.53}$$

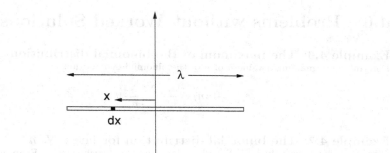

$$\text{moment of inertia} = 2\int_0^{\lambda/2} x^2 \, dx \, (M/\lambda) = M(\lambda/2)^2/3$$

Fig. 4.4 The moment of inertia of the needle about the axis through its centre.

Correspondingly we obtain for the mean-square angular velocity:

$$\overline{\dot{\theta}^2} = \frac{\int_0^\infty L^2 e^{-L^2/2IkT}\,dL}{I^2 \int_0^\infty e^{-L^2/2IkT}\,dL} \overset{(2.20)}{=} \frac{1}{I^2}\frac{\sqrt{\pi(2IkT)^3}/4}{\sqrt{\pi 2IkT}/2} = \frac{kT}{I} = \frac{3kT}{M(\lambda/2)^2}. \tag{4.54}$$

Thus finally:

$$\sqrt{\overline{\dot{\theta}^2}} = \frac{2}{\lambda}\sqrt{\frac{3kT}{M}}. \tag{4.55}$$

Example 4.5: Root-mean-square values by equipartition of energy
Show that the results of Example 4.4 can be derived from the law of equipartition of energy.

Solution: We obtain from the preceding problem the energy of the needle as

$$\epsilon = \frac{1}{2}I\dot{\theta}^2 + \frac{1}{2}a\theta^2 = \frac{L^2}{2I} + \frac{1}{2}a\theta^2, \quad \text{where} \quad L = I\dot{\theta}. \tag{4.56}$$

According to the law of equipartition of energy (4.26b) every quadratic form in the energy contributes a mean energy of $kT/2$. Thus in the present case we obtain for the mean-square angular deflection of the needle:

$$\frac{kT}{2} = \frac{1}{2}a\overline{\theta^2}, \tag{4.57}$$

from which we immediately deduce that

$$\sqrt{\overline{\theta^2}} = \sqrt{\frac{kT}{a}}. \tag{4.58}$$

Similarly we obtain from the kinetic term in the energy:

$$\frac{kT}{2} = \frac{1}{2}I\overline{\dot{\theta}^2}, \tag{4.59}$$

so that

$$\sqrt{\overline{\dot{\theta}^2}} = \sqrt{\frac{kT}{I}} = \frac{2}{\lambda}\sqrt{\frac{3kT}{M}}. \tag{4.60}$$

These results are seen to agree with those of Example 4.4.

4.6   Problems without Worked Solutions

Example 4.6: The maximum of the binomial distribution
Determine the maximum value $\tilde{n}$ of n of the binomial distribution

$$W(n) = \frac{N!}{n!(N-n)!}p^n q^{N-n}. \tag{4.61}$$

Example 4.7: The binomial distribution for large N, n
Show that for large values of N and n the binomial distribution of Example 4.6 assumes the Gaussian form

$$W(n) = Ne^{-|V_2|(n-\tilde{n})^2/2}, \tag{4.62}$$

and determine the normalization constant N by evaluating

$$\sum_{n=0}^{N} W(n) \simeq \int_0^N W(n)dn = 1 \tag{4.63}$$

(Answer: $N = \sqrt{|V_2|/2\pi}$).

Example 4.8: The Poisson distribution
Show that with the help of the Stirling approximation (and $n \ll N, p \ll 1$) the binomial distribution $W(n)$ of Example 4.6 with $q = 1 - p$ reduces to the following expression known as the Poisson distribution, *i.e.*

$$W(n) \simeq \frac{\lambda^n}{n!}e^{-\lambda}, \ \lambda = Np. \tag{4.64}$$

Show that the result can also be obtained with the help of the formula

$$e^{-\lambda} = \lim_{n \to \infty}\left(1 - \frac{\lambda}{n}\right)^n.$$

Example 4.9: Probability of sedimentation of atoms
N atoms of the same kind in exhaust fumes of some industrial complex are blown by wind into the air and deposit on elements of area a of a larger surrounding region of area A. On the average $\dot{N} = dN/dt$ atoms are deposited per minute on an element a. Show that the probability

of sedimentation of n atoms on an element of area a in a time interval t is given by the Poisson distribution

$$W(n) = \frac{\lambda^n}{n!} e^{-\lambda}, \tag{4.65}$$

where $\lambda = Np$, and p is the probability a/A.

Example 4.10: Decrease of energy fluctuation with increasing N
(a) It is a typical characteristic of statistical distributions that, for instance, the percentage deviation of the total energy E of an N-particle system from its mean value $\overline{E}$ becomes increasingly negligible with rapidly increasing N. Verify this statement by verification of the relation

$$D(E) = \frac{1}{\sqrt{N}} D(\epsilon), \quad \text{where} \quad D^2(E) = \frac{\overline{(E - \overline{E})^2}}{(\overline{E})^2} = \frac{\overline{E^2} - \overline{E}^2}{(\overline{E})^2}, \tag{4.66}$$

i.e. show that

$$D^2(E) = \frac{1}{N} \frac{\overline{\epsilon^2} - \overline{\epsilon}^2}{(\overline{\epsilon})^2} = \frac{1}{N} D^2(\epsilon), \tag{4.67}$$

where $\overline{\epsilon} \equiv \overline{\epsilon_i}$ and ϵ_i is the energy of the i-th particle (classical statistics, take $n_i = 1$, $E = \sum \epsilon_i$).
(b) Evaluate $D^2(\epsilon)$ in classical statistics of noninteracting, nonrelativistic particles in a volume V at absolute temperature T. (The answer follows from $\overline{\epsilon_i} = 3kT/2$, $\overline{\epsilon_i^2} = 15(kT)^2/4$.)

Example 4.11: Partition of energy weighting factor of diatomic molecule
Consider the energy ϵ of a diatomic molecule as the sum of three independent contributions ϵ_{trans}, ϵ_{rot} and ϵ_{vib} arising from translation, rotation and vibration. Consider the relative motion of one of the atoms about the other as if one of them had an infinite mass which acts as a center for the other with reduced mass $m_0 = m_1 m_2/(m_1 + m_2)$. Then, using polar coordinates as in Example 3.6, with $p_\theta = m_0 r^2 \dot\theta$ and $p_\phi = m_0 r^2 \dot\phi \sin^2 \theta$, the relevant energy is

$$\epsilon = \frac{1}{2} m_0 \dot{r}^2 + \frac{1}{2} m_0 r^2 \dot\theta^2 + \frac{1}{2} r^2 \dot\phi^2 \sin^2 \theta + \epsilon_{vib} = \frac{p_r^2}{2m_0} + \frac{p_\theta^2}{2I} + \frac{p_\phi^2}{2I \sin^2 \theta} + \epsilon_{vib}, \tag{4.68}$$

where I is the moment of inertia of the molecule. Approximating ϵ_{vib} by an harmonic potential about an equilibrium position r_0, i.e. setting

$$\epsilon_{trans} + \epsilon_{vib} = \frac{p_r^2}{2m_0} + \frac{1}{2a^2}(r - r_0)^2, \quad a = \text{const.}, \tag{4.69}$$

and quantizing this in the usual way, we can set

$$\epsilon_{trans} + \epsilon_{vib} = h\nu\left(n + \frac{1}{2}\right), \quad n = 0, 1, 2, \dots, \quad \text{and} \quad \nu = \frac{1}{2\pi a\sqrt{m_0}}. \tag{4.70}$$

With ϵ_{rot} also quantized we have

$$\epsilon = \frac{J(J+1)\hbar^2}{2I} + h\nu\left(n + \frac{1}{2}\right). \tag{4.71}$$

(a) Show that the vibrational part of the internal partition function is

$$\sum e^{-\epsilon_{vib}/kT} = \sum_{n=0}^{\infty} e^{-(n+1/2)h\nu/kT} = \frac{\exp(-h\nu/2kT)\exp(h\nu/kT)}{\exp(h\nu/kT) - 1}. \tag{4.72}$$

Explain why the factor after the sum is 1.

(b) Show that the rotational part of the internal partition function is

$$\sum e^{-\epsilon_{\text{rot}}/kT} = \sum_{J=0}^{\infty}(2J+1)e^{-J(J+1)h^2/2IkT} = 8\pi^2 IkT. \tag{4.73}$$

Show that the same result is obtained from the classical expression, *i.e.* from

$$\int d\theta d\phi dp_\theta dp_\phi \exp\left\{-\frac{p_\theta^2}{2IkT} - \frac{p_\phi^2}{2I\sin^2\theta kT}\right\}. \tag{4.74}$$

Example 4.12: Mean length of an elastic string
An elastic string with a mass M at one end is hanging in the gravitational field of the earth. The string is to be considered as consisting of N equal elements, each of fixed length a, such that these can move up or down with no other forces of significance. Show that at temperature T the mean length $\overline{L}$ of the string is given by

$$\overline{L} = Na\tanh\left(\frac{Mga}{kT}\right). \tag{4.75}$$

Explain the result in the limit $T=0$.

Example 4.13: Why doesn't the atmosphere fall down?
Calculate the average value of the potential energy of a molecule of mass m in an isothermal atmosphere. How do you explain the result in contrast with the law of equipartition of energy?[¶¶]

[¶¶]For a discussion see A.F. Brown [7], pp.75 and 145.

Chapter 5

Entropy

5.1 Introductory Remarks

What is this thing called entropy? asks Ben–Naim in his book *Entropy Demystified*,[*] and pinpoints the turning point in his own understanding of entropy on his realization that the two key features of the atomic theory of matter are (a) the huge numbers involved, and (b) the indistinguishability of the particles constituting matter.[†] Only on the basis of the atomic theory of matter can entropy be understood and this is as given by Boltzmann's formula. This formula is the subject of this chapter, however here in the context of (classical) Maxwell–Boltzmann statistics which treats the particles as distinguishable (the indistinguishability will appear in quantum statistics in Chapter 6).

5.2 The Boltzmann Formula

Consider the logarithm of the number of arrangements W_{MB} of elements in Maxwell–Boltzmann statistics, *i.e.* $\ln W_{MB}$, given by Eq. (4.2). We make the following observations:

1. $\ln W_{MB}$ depends only on the present state of the system, but not on the path of arrival to that state.

2. We have

$$W_{MB} = W_{MB}^{(1)} \times W_{MB}^{(2)} \times \cdots ,$$

---

[*]A. Ben–Naim [4], p.x.
[†]A. Ben–Naim [4], p.xiii.

where $W_{MB}^{(i)} \propto$ probability of the number of arrangements of elements of system i when the systems are in contact. Therefore

$$\ln W_{MB} = \ln W_{MB}^{(1)} + \ln W_{MB}^{(2)} + \cdots .$$

We see therefore that $\ln W_{MB}$ is additive for systems in contact (and also when the systems are not in contact).

3. The expression $\ln W_{MB}$ tends to a maximum, but not to infinity because of subsidiary conditions.

The listed properties are three of the most important properties of the *thermodynamical entropy* S. Is the entropy S proportional to $\ln W_{MB}$? We now examine the properties of $\ln W_{MB}$ more closely.

For a classical system we have the Maxwell–Boltzmann number of arrangements as in Eq. (4.2),

$$W_{MB} = N! \prod_i \frac{g_i^{n_i}}{n_i!}. \tag{5.1}$$

Hence, using *Stirling's theorem*, Eq. (4.5),

$$\ln W_{MB} \simeq N(\ln N - 1) + \sum_i [n_i \ln g_i - n_i (\ln n_i - 1)]. \tag{5.2}$$

But for the classical system (which is a conserved system)

$$N = \sum_i n_i, \tag{5.3}$$

so that two terms cancel and we have

$$\ln W_{MB} \simeq N \ln N + \sum_i [n_i \ln g_i - n_i \ln n_i]$$

$$= N \ln N - \sum_i n_i \ln \left(\frac{n_i}{g_i} \right). \tag{5.4}$$

However, we had, by maximizing $\ln W_{MB}$, Eq. (4.14) (with $\mu = 1/kT$ there), *i.e.*

$$n_i = N \frac{g_i e^{-\epsilon_i/kT}}{\sum_i g_i e^{-\epsilon_i/kT}}, \tag{5.5}$$

where the expression in the denominator is the normalization factor for conserved elements. It follows that

$$\frac{n_i}{g_i} = \frac{N e^{-\epsilon_i/kT}}{\sum_j g_j e^{-\epsilon_j/kT}}. \tag{5.6}$$

Thus in the *state of maximum probability*

$$\ln W_{MB} = N \ln N - \sum_i n_i \left[\ln N - \frac{\epsilon_i}{kT} - \ln \sum_j g_j e^{-\epsilon_j/kT} \right]$$

$$= \frac{1}{kT} \sum_i n_i \epsilon_i + N \ln \sum_i g_i e^{-\epsilon_i/kT}. \tag{5.7}$$

But

$$\sum_i n_i \epsilon_i = E = \text{ total energy.} \tag{5.8}$$

Therefore

$$k \ln W_{MB} = \frac{E}{T} + Nk \ln \sum_i g_i e^{-\epsilon_i/kT}. \tag{5.9}$$

Here the second term on the right is independent of E. From thermodynamics[‡] we know that the quantity known as the *free energy* F of the system (also called Helmholtz free energy) is defined by the *Gibbs–Helmholtz equation* (strictly speaking F, E, S are averages)[§]

$$F = E - TS, \quad S = \frac{E}{T} - \frac{F}{T}. \tag{5.10}$$

Therefore we obtain the very famous *Boltzmann formula* (inscribed on his tombstone)

$$S \to S_{MB} = k \ln W_{MB}, \tag{5.11}$$

along with the relations

$$F \to F_{MB} = -NkT \ln \sum_i g_i e^{-\epsilon_i/kT} \equiv -kT \ln Z, \tag{5.12}$$

and

$$Z = \left(\sum_i g_i e^{-\epsilon_i/kT} \right)^N. \tag{5.13}$$

Observe here that according to our original definition of g_i, this g_i appearing here contains also the volume (say) V. The function Z thus defined is called the *partition function of the system* (here with power N the N-particle partition function), and is a function of T, V and N. The expression for Z will be obtained as an exact result (without any approximation) in Chapter 12,

---

[‡]We mean ordinary, classical macroscopic thermodynamics as *e.g.* in the book of H.C. Callen [10].

[§]Thus, since F is *defined* as $\overline{E} - T\overline{S}$, where $\overline{E}$ and $\overline{S}$ are averages, there is really no need to write F as $\overline{F}$. This is a matter of definition.

where the result is Eq. (12.18). We shall see in Sec. 6.5 that the classical Maxwell–Boltzmann expression for the free energy F in Eq. (5.12) is wrong by a factor $N!$. Note that the partition function is a measure of probability. Thus in the case of a system with rotational, vibrational and electronic degrees of freedom, and the energy ϵ_i of an element (molecule) being the sum $\epsilon^{\text{rot}} + \epsilon^{\text{vib}} + \epsilon^{\text{elec}}$, the corresponding partition functions are multiplied, *i.e.*

$$Z = Z^{\text{rot}} Z^{\text{vib}} Z^{\text{elec}}, \qquad Z^{\text{rot}} = \left(\sum_i g_i^{\text{rot}} e^{-\epsilon_i^{\text{rot}}/kT} \right)^N, \text{ etc.} \qquad (5.14)$$

Alternatively, starting from the *second law of thermodynamics*,[¶] we have

$$dQ = dE + PdV, \quad i.e. \quad TdS = dE + PdV. \qquad (5.15)$$

It follows that

$$dS = \frac{1}{T}dE + \frac{P}{T}dV. \qquad (5.16)$$

From this relation we deduce that

$$\frac{1}{T} = \left(\frac{\partial S}{\partial E} \right)_V, \quad \text{and} \quad \frac{P}{T} = \left(\frac{\partial S}{\partial V} \right)_E \qquad (5.17)$$

(here the second relation is of little importance). But with Eq. (5.9):

$$\left[\frac{\partial (k \ln W_{MB})}{\partial E} \right]_V = \frac{1}{T}, \qquad (5.18)$$

and therefore we obtain the important formula:

$$S_{MB} = k \ln W_{MB} \overset{(5.9)}{=} \frac{E}{T} + k \ln Z. \qquad (5.19)$$

Here we note that the relation (5.18) provides us with the *statistical definition* of the *absolute temperature* T, which we introduced in this text originally through the phenomenologically observed gas law in Sec. 2.4.1.[‖] Now F follows from $E - TS = F$ as the following very important expression:

$$F_{MB} = -NkT \ln \sum_i g_i e^{-\epsilon_i/kT} = -kT \ln Z. \qquad (5.20)$$

From this vital equation everything else can be derived in classical statistics! For example, since the Gibbs–Helmholtz equation implies

$$dF \overset{(1.11)}{=} dE - TdS - SdT \overset{(5.15)}{=} -PdV - SdT, \qquad (5.21)$$

[¶]Here the macroscopic quantities E, P, S, Q are really *average* values, and would more properly be written $\overline{E}, \overline{P}, \overline{S}, \overline{Q}$. We assume the reader keeps this in mind.

[‖]For considerable discussion we refer to F. Reif [59], Sec. 3.5.

we obtain S from the relation

$$S = -\left(\frac{\partial F}{\partial T}\right)_V = \frac{E - F}{T}, \tag{5.22}$$

and E from Eqs. (5.10) and (5.12) as the following expression as shown in Example 5.1:

$$E = -T^2\left[\frac{\partial (F/T)}{\partial T}\right]_V = -T^2\frac{\partial (F/T)}{\partial (1/T)}\frac{\partial (1/T)}{\partial T}$$

$$\overset{(5.20)}{=} -\frac{\partial \ln Z}{\partial \beta}, \quad \beta = \frac{1}{kT}. \tag{5.23}$$

Thus the energy, more precisely the macroscopic mean energy $\overline{E}$, follows from the important relation[**]

$$\overline{E} = -\frac{\partial \ln Z}{\partial \beta}.$$

Further we obtain from the *Gibbs–Helmholtz equation* the specific heat C_V at constant volume V in terms of the free energy F, *i.e.* from Eqs. (5.15) and (1.21):

$$C_V = \left(\frac{\partial E}{\partial T}\right)_V, \tag{5.24}$$

or

$$C_V \overset{(5.15)}{=} T\left(\frac{\partial S}{\partial T}\right)_V \overset{(5.22)}{=} -T\left(\frac{\partial^2 F}{\partial T^2}\right). \tag{5.25}$$

We obtain the pressure P from $dF = -PdV - SdT$ of Eq. (5.21) as

$$P = -\left(\frac{\partial F}{\partial V}\right)_T. \tag{5.26}$$

With these relations permitting derivations from the important *free energy* F we end our consideration of classical statistics.

5.3 Applications and Examples

Example 5.1: Derivation of entropy and energy from the free energy
Assuming that the entropy S and the free energy F are given by Eqs. (5.10) to (5.12),

$$S = k \ln W = \frac{E}{T} + Nk \ln \sum_i g_i e^{-\epsilon_i/kT} \quad \text{and} \quad F = E - TS = -NkT \ln \sum_i g_i e^{-\epsilon_i/kT}, \tag{5.27}$$

[**]The corresponding relation in the ensemble consideration of Chapter 11 is therefore Eq. (11.13).

verify (here in classical statistics) that

$$S = -\left(\frac{\partial F}{\partial T}\right)_V \quad \text{and} \quad E = -T^2\left[\frac{\partial}{\partial T}\left(\frac{F}{T}\right)\right]_V. \tag{5.28}$$

Solution: We recall the second law of thermodynamics and the definition of the free energy F. These are given by the relations

$$dE(S,V) = TdS - PdV, \quad F(T,V) = E(S,V) - TS, \quad dF \stackrel{(1.11)}{=} -SdT - PdV. \tag{5.29}$$

We see from these relations that

$$S = -\left(\frac{\partial F}{\partial T}\right)_V. \tag{5.30}$$

We obtain $\partial F/\partial T$ in explicit form as follows. Differentiating Eq. (5.20) with respect to T we have here in the Maxwell–Boltzmann case:

$$
\begin{aligned}
\frac{\partial F_{MB}}{\partial T} &= -Nk\ln\sum_i g_i e^{-\epsilon_i/kT} - NkT\frac{\partial}{\partial T}\ln\sum_i g_i e^{-\epsilon_i/kT} \\
&= -Nk\ln\sum_i g_i e^{-\epsilon_i/kT} - NkT\frac{\sum_i g_i(\epsilon_i/kT^2)e^{-\epsilon_i/kT}}{\sum_i g_i e^{-\epsilon_i/kT}} \\
&= -Nk\ln\sum_i g_i e^{-\epsilon_i/kT} - \frac{\sum_i \epsilon_i(Ng_i e^{-\epsilon_i/kT})}{T\sum_i g_i e^{-\epsilon_i/kT}} \\
&\stackrel{(5.6)}{=} -Nk\ln\sum_i g_i e^{-\epsilon_i/kT} - \frac{\sum_i \epsilon_i n_i \sum_j g_j e^{-\epsilon_j/kT}}{T\sum_i g_i e^{-\epsilon_i/kT}} \\
&= -Nk\ln\sum_i g_i e^{-\epsilon_i/kT} - \frac{E}{T} = \frac{F_{MB}-E}{T} = -S_{MB}.
\end{aligned} \tag{5.31}
$$

For the second part we consider

$$\frac{F_{MB}}{T} = -Nk\ln\sum_i g_i e^{-\epsilon_i/kT}. \tag{5.32}$$

Differentiating with respect to T we obtain:

$$
\begin{aligned}
\frac{\partial}{\partial T}\left(\frac{F_{MB}}{T}\right) &= -Nk\frac{\sum_i g_i(\epsilon_i/kT^2)e^{-\epsilon_i/kT}}{\sum_i g_i e^{-\epsilon_i/kT}} \stackrel{(5.6)}{=} -\frac{\sum_i \epsilon_i n_i \sum_j g_j e^{-\epsilon_j/kT}}{T^2\sum_i g_i e^{-\epsilon_i/kT}} \\
&= -\frac{\sum_i n_i\epsilon_i \sum_j g_j e^{-\epsilon_j/kT}}{T^2\sum_i g_i e^{-\epsilon_i/kT}} = -\frac{E}{T^2}.
\end{aligned} \tag{5.33}
$$

It follows that

$$E = -T^2\left[\frac{\partial}{\partial T}\left(\frac{F_{MB}}{T}\right)\right]_V. \tag{5.34}$$

Example 5.2: Mean speeds and root-mean-square speeds

Show that in a Maxwell–Boltzmann gas the number of molecules having speeds between c and $c + dc$ is

$$dN = \frac{4Nc^2 e^{-mc^2/2kT}}{\sqrt{\pi}(2kT/m)^{3/2}}dc. \tag{5.35}$$

Show that the mean speed $\bar{c}$ and root-mean-square speed $\sqrt{\bar{c^2}}$ are given by

$$\bar{c} = \frac{2c_m}{\sqrt{\pi}} \quad \text{and} \quad \sqrt{\bar{c^2}} = \sqrt{\frac{3}{2}}c_m, \tag{5.36}$$

where c_m is the most probable speed. How does c_m vary with m and T?

Solution: We obtain the first result with Eq. (5.5), *i.e.*

$$n_i = N\frac{g_i e^{-\epsilon_i/kT}}{\sum_i g_i e^{-\epsilon_i/kT}} \equiv dN. \tag{5.37}$$

With Eq. (3.9c) we have here $g_i = V4\pi p^2 dp/h^3 = (V/h^3)4\pi m^3 c^2 dc$ and

$$\sum_i g_i e^{-\epsilon_i/kT} = \frac{V}{h^3}4\pi m^3 \int_0^\infty e^{-mc^2/2kT}c^2 dc \stackrel{(2.20)}{=} \frac{4\pi V}{h^3}m^3\frac{1}{4}\sqrt{\frac{\pi(2kT)^3}{m^3}}. \tag{5.38}$$

Hence

$$dN = N\frac{(V/h^3)4\pi m^3 c^2 dc\, e^{-mc^2/2kT}}{(V/h^3)4\pi m^3(1/4)\sqrt{\pi(2kT)^3/m^3}} = \frac{4c^2 dc\, e^{-mc^2/2kT}}{\sqrt{\pi}(2kT/m)^{3/2}}N. \tag{5.39}$$

The remaining parts of the problem follow as in Examples 2.5 and 2.6. Thus from there we have

$$c_m \stackrel{(2.60)}{=} \sqrt{\frac{2kT}{m}}, \quad \bar{c} \stackrel{(2.55)}{=} 2\sqrt{\frac{2kT}{m\pi}} = \frac{2}{\sqrt{\pi}}c_m, \quad \bar{c^2} \stackrel{(2.57)}{=} \frac{3kT}{m} = \frac{3}{2}c_m^2. \tag{5.40}$$

Example 5.3: Thermodynamics of N magnetic moments

Examine the magnetic contributions to the thermodynamical functions of Maxwell–Boltzmann statistics for a system of N magnetic moments in a uniform magnetic field.

Solution: We recall that g_i contains the dependence on the potential, like (say) V. Thus we have in the present case (*cf.* Example 3.2, Eq. (3.17)):

$$g_i = \text{constant}\, V \sin\theta\, d\theta. \tag{5.41}$$

The entropy S, here S_{MB}, is given by $S_{MB} = k\ln W_{MB}$, and the free energy F, here F_{MB}, by Eq. (5.20),

$$F_{MB} = -NkT\ln\left(\sum_i g_i e^{-\epsilon_i/kT}\right). \tag{5.42}$$

Inserting here the energy of the magnetic moment as in Example 4.3, we have

$$\begin{aligned}
F_{MB} &= -NkT\ln\left(\text{const.}V\int_0^\pi \sin\theta\, d\theta\, e^{MH\cos\theta/kT}\right) \\
&= -NkT\ln\left[\text{const.}V\frac{e^{MH\cos\theta/kT}}{-MH/kT}\right]_0^\pi = -NkT\ln\left[\frac{\text{const.}V}{MH/kT}\left(e^{MH/kT} - e^{-MH/kT}\right)\right] \\
&= -NkT\ln\left[\frac{2kT}{MH}\text{const.}V\sinh\left(\frac{MH}{kT}\right)\right].
\end{aligned} \tag{5.43}$$

This is the contribution to the free energy of a system resulting from the N magnetic moments in the magnetic field.[††] We obtain its contribution to the pressure P of the system from formula (5.26) and hence obtain

$$P = -\left(\frac{\partial F}{\partial V}\right)_T = \frac{NkT}{V}. \tag{5.44}$$

[††]This is the result frequently given in textbooks, *e.g.* in S.K. Ma [42], where (p.278) the result is given with the remark: "The free energy can be calculated very easily."

We recognize this result as the gas law (*i.e.* of a perfect gas). Thus the N magnetic moments contribute to the pressure like N molecules of a perfect gas. We observe that if we differentiate $-F$ with respect to the magnetic field strength H, we reproduce here for N elements the mean magnetic moment (4.46) of Example 4.3, *i.e.* we have:

$$
-\frac{\partial F_{MB}}{\partial H}
$$

$$
= NkT \frac{(2kT/MH)\cosh(MH/kT)V\text{const.}(M/kT) - 2kT/MH^2\text{const.}V\sinh(MH/kT)}{(2kT/MH)\text{const.}V\sinh(MH/kT)}
$$

$$
= NkT \frac{(2kT/MH)\cosh(MH/kT).(M^2H/2(kT)^2) - (MH/2kT)(2kT/MH^2)\sinh(MH/kT)}{\sinh(MH/kT)}
$$

$$
= NkT[(2kT/MH)\coth(MH/kT).(M^2H/2(kT)^2) - (1/H)]
$$

$$
= NM\left[\coth\left(\frac{MH}{kT}\right) - \frac{kT}{HM}\right], \quad \text{so that} \quad N\overline{M} \overset{(4.46)}{=} -\frac{\partial F_{MB}}{\partial H}. \tag{5.45}
$$

This well-known result was obtained by Langevin in 1905 and can be found (derived) in many books (see, *e.g.*, Allis and Herlin [1], p.177).

Example 5.4: Free energy, latent heat on Maxwell–Boltzmann statistics

Calculate the free energy on Maxwell–Boltzmann statistics,

$$
F_{MB} = -NkT\ln\left[\sum_i g_i e^{-\epsilon_i/kT}\right], \tag{5.46}
$$

of a system of N free atoms in a volume V. Show that with these statistics the entropy S is

$$
S_{MB} = -\left(\frac{\partial F_{MB}}{\partial T}\right)_V = Nk\ln V + \frac{3}{2}Nk\ln T + Nk\ln\left(\frac{2\pi mk}{h^2}\right)^{3/2} + \frac{3}{2}Nk. \tag{5.47}
$$

Calculate the latent heat of evaporation per atom — which one would expect to be independent of the number of atoms, at a given temperature — and show that the resultant expression is not acceptable, because it would imply that the volume of vapour at a given temperature would be independent of the number of atoms in the volume.

Solution: We have the *a priori* probability

$$
g_i = \frac{V.4\pi p^2 dp}{h^3}. \tag{5.48}
$$

$$
\therefore \sum_i g_i e^{-\epsilon_i/kT} = \frac{4\pi V}{h^3}\int_0^\infty p^2 e^{-p^2/2mkT}dp \overset{(2.20)}{=} \frac{4\pi V}{4h^3}\sqrt{\pi(2mkT)^3}
$$

$$
= \frac{V}{h^3}(2\pi mkT)^{3/2} = V\left(\frac{2\pi mkT}{h^2}\right)^{3/2}. \tag{5.49}
$$

It follows that

$$
F_{MB} = -NkT\ln\left[V\left(\frac{2\pi mkT}{h^2}\right)^{3/2}\right] = -NkT\left[\ln V + \ln\left(\frac{2\pi mk}{h^2}\right)^{3/2} + \frac{3}{2}\ln T\right]. \tag{5.50}
$$

Thus with this result we obtain

$$
-\left(\frac{\partial F_{MB}}{\partial T}\right)_V = Nk\ln V + Nk\ln\left(\frac{2\pi mk}{h^2}\right)^{3/2} + \frac{3}{2}Nk\ln T + \frac{3}{2}Nk. \tag{5.51}
$$

We have for the latent heat of the vapour:

$$L_{\text{vapour}} = \frac{1}{NT}\left(S_{MB \text{ vapour}} - S_{MB \text{ liquid}}\right) \text{ per atom,} \tag{5.52}$$

where $S_{MB \text{ liquid}}$ is negligible. Thus

$$L_{\text{vapour}} = \frac{k}{T}\ln V + \frac{k}{T}\ln\left(\frac{2\pi mk}{h^2}\right)^{3/2} + \frac{3}{2}\frac{k}{T}\ln T + \frac{3}{2}\frac{k}{T} \equiv f(T, V). \tag{5.53}$$

Since at constant temperature T, L_{vapour} is constant (and one expects this on physical grounds to be independent of the number of atoms N), the volume V (in $\ln V$) of the vapour at a given temperature would according to Eq. (5.53) be independent of the number of atoms in the volume. Hence this expression is not acceptable but we shall see in Sec. 6.6 and Example 6.7 how this result is corrected in quantum statistics.

Example 5.5: Calculation of pressure

Using S_{MB} and F_{MB}, calculate the pressure P of a perfect gas from the relations

$$P \overset{(5.17)}{=} T\left(\frac{\partial S}{\partial V}\right)_E, \qquad P \overset{(5.26)}{=} -\left(\frac{\partial F}{\partial V}\right)_T. \tag{5.54}$$

Solution: From Example 3.1 we know that in the case of volume-independent energies ϵ_i (like kinetic energies) the degeneracy g_i is proportional to the volume V. In general the eigenvalues ϵ_i depend on the volume V, and hence also the total energy $E = \sum_i n_i\epsilon_i$. Thus more generally, in differentiating with respect to V at constant T the relation

$$S_{MB} \overset{(5.19)}{=} k \ln W_{MB} = \frac{E}{T} + Nk\ln\sum_i g_i e^{-\epsilon_i/kT}, \tag{5.55}$$

we have also the two terms (*i.e.* excluding differentiation of g_i)

$$\frac{1}{T}\frac{\partial E}{\partial V} - \frac{Nk}{kT}\frac{\sum_i g_i(\partial\epsilon_i/\partial V)e^{-\epsilon_i/kT}}{\sum_i g_i e^{-\epsilon_i/kT}} \overset{(5.5)}{=} \frac{1}{T}\frac{\partial E}{\partial V} - \frac{1}{T}\sum_i n_i\frac{\partial\epsilon_i}{\partial V} = \frac{1}{T}\frac{\partial E}{\partial V} - \frac{1}{T}\frac{\partial}{\partial V}\sum_i n_i\epsilon_i = 0. \tag{5.56}$$

Thus we only have to differentiate in Eq. (5.55) the factor g_i, and we have with this one remaining term (multiplying the denominator by V and differentiating g_i there with respect to V)

$$\frac{P}{T} = \left(\frac{\partial S_{MB}}{\partial V}\right)_E = \frac{Nk}{V}\frac{\sum_i(\partial g_i/\partial V)e^{-\epsilon_i/kT}}{\sum_i(\partial g_i/\partial V)e^{-\epsilon_i/kT}} = \frac{Nk}{V}, \quad \text{implying} \quad PV = NkT. \tag{5.57}$$

We thus obtain the law of a perfect gas. On the other hand, using the free energy F_{MB} with energies ϵ_i independent of V, we start from

$$F_{MB} = -NkT\ln\left[\sum_i g_i e^{-\epsilon_i/kT}\right], \tag{5.58}$$

so that the pressure P is given by (with $F \to F_{MB}$)

$$P = -\left(\frac{\partial F}{\partial V}\right)_T = NkT\frac{\sum_i(\partial g_i/\partial V)e^{-\epsilon_i/kT}}{\sum_i g_i e^{-\epsilon_i/kT}} = NkT\frac{\sum_i(\partial g_i/\partial V)e^{-\epsilon_i/kT}}{V\sum_i(\partial g_i/\partial V)e^{-\epsilon_i/kT}}$$

$$= \frac{NkT}{V}, \quad \text{so that} \quad PV = NkT. \tag{5.59}$$

From Eq. (4.27b) we obtain as the energy E of N noninteracting particles the expression (which is constant at constant T)

$$E = N\epsilon = N\frac{3}{2}kT, \qquad i.e. \quad PV = NkT = \frac{2}{3}E. \tag{5.60}$$

(See also Example 7.3).

5.4   Problems without Worked Solutions

Example 5.6: Maximum of entropy
The entropy S (here S_{MB}) and the canonical partition function Z are related as *e.g.* in Eq. (5.19) which can be rewritten in the form (with $g_i = 1$)

$$S = k(\ln Z + \overline{E}/kT), \quad Z = \sum_i e^{-\epsilon_i/kT}, \quad P_i = \frac{1}{Z}e^{-\epsilon_i/kT}. \tag{5.61}$$

(a) Show that the probability P_i and S are connected by the following relation also called the Gibbs entropy [12]:

$$S = -k\sum_i P_i \ln P_i, \quad \sum_{i=1}^N P_i = 1. \tag{5.62}$$

This formula describes the thermal entropy S of the second law of thermodynamics. P_i is the Boltzmann factor which represents the probability to find the system excited to its state i in a heat bath with temperature T with which it is in a state of equilibrium. In quantum mechanical operator form for a system with Hamiltonian H the entropy is described by the operator

$$\rho_{\mathrm{MB}} = \frac{e^{-H/kT}}{Tre^{-H/kT}}. \tag{5.63}$$

This expression is called the Maxwell–Boltzmann density matrix. The entropy is then given by

$$S = -\mathrm{Tr}\rho_{\mathrm{MB}} \ln \rho_{\mathrm{MB}}. \tag{5.64}$$

(b) A certain system can occupy one of N states. The expression P_i is the probability for the system to occupy state i ($i = 1, \ldots, N$). Show with the help of the result of (a) that the maximum entropy is given by

$$S = k \ln N. \tag{5.65}$$

Note the difference between cases (a) and (b). The result (5.65) is that for the maximized entropy (different from (5.62)). Compare also with Eq. (5.11).
[Hint: Use the method of Lagrangian multipliers to link $\delta S, \delta P_i$].

(c) The accessible states of a particular system can be subdivided into W sets, the i-th set consisting of G_i states of equal probabilities. The probability for the system to be in any one of these (equally probable) states is P_i. What is in this case the expression for the entropy?

Example 5.7: Density matrix equation
Show with the help of the Schrödinger equation

$$i\hbar \frac{\partial}{\partial t}\psi(\mathbf{x}, t) = H\psi(\mathbf{x}, t), \tag{5.66}$$

that the density operator ρ_N, *i.e.*

$$\rho_N(\beta) = e^{-\beta H} \tag{5.67}$$

(without normalization), satisfies in the configuration space matrix representation the equation

$$\frac{\partial \rho_N(x, x'; \beta)}{\partial \beta} = -H_x \rho_N(x, x'; \beta), \tag{5.68}$$

where the subscript x indicates that H acts on x of $\rho_N(x, x'; \beta)$.[‡‡] Note: Observe that the Schrödinger equation describes a single, so-called pure state. The thermal density equation, however, with temperature dependence ($\beta = 1/kT$) describes the overall state of a system with mixed states. The equation can also be found with the names of Liouville and/or von Neumann attached.

[‡‡]See *e.g.* H.J.W. Müller-Kirsten [47], Sec. 5.4.1.

Example 5.8: Mean square deviation of energy

Show that the mean square deviation of the energy E, *i.e.*

$$\overline{(\Delta E)^2} = \overline{(E - \overline{E})^2} = \overline{E^2} - \overline{E}^2, \tag{5.69}$$

is given by

$$\overline{(\Delta E)^2} = \frac{\partial^2 \ln Z}{\partial \beta^2} = -\frac{\partial \overline{E}}{\partial \beta}, \qquad \beta = \frac{1}{kT}. \tag{5.70}$$

Example 5.9: Relative square of fluctuation of pressure

Using the result (5.60), *i.e.* $PV = 2E/3$, show that the relative square of the fluctuation of the pressure P of an ideal gas of N particles is given by

$$\frac{\overline{(\Delta P)^2}}{\overline{P}^2} = \frac{2}{3N}. \tag{5.71}$$

What do you conclude from this result for large N?

Example 5.10: Entropy of a bit

What is the entropy of a bit?
(Answer: $k \ln 2$).

Example 5.11: Entropy of a phase space volume V

Show that if the probability density in phase space $\rho(p, q)$ is $\rho = 1/V$ inside a region Γ and zero outside, then the entropy is $S = \ln V$.

Example 5.12: Entropy of a system of spins

Show that the maximum entropy of a lattice of discrete spins (lattice spacing a, volume V) is $V \ln 2/a^3$.

Example 5.13: Negative specific heat

Ordinarily the specific heat is positive and raises the temperature of a system. What happens if the specific heat happened to be negative (as in the case of a black hole)?

Chapter 6

Quantum Statistics

6.1 Introductory Remarks

There are two main differences between *quantum statistics* and *classical statistics*:

1. In quantum statistics the energy levels are nearly always discrete. In classical statistics they are assumed to be continuous. Consider for instance the simple or one-dimensional linear harmonic oscillator. In quantum mechanics its (nondegenerate) energy levels for one element of the system are given by

$$\epsilon_n = h\nu\left(n + \frac{1}{2}\right), \quad n = 0, 1, 2, \ldots . \tag{6.1}$$

   In classical statistics it has the energy

$$\epsilon = \frac{1}{2m}p^2 + 2\pi^2 m\nu^2 q^2. \tag{6.2}$$

2. The counting of the number of arrangements of the elements is different in the two types of statistics. A dominant reason for this is the *indistinguishability of the elements in quantum mechanics* (resulting directly from the uncertainty relation). This is by far the most important effect.

The difference between particles in a box in classical mechanics and those in a box in quantum mechanics is illustrated by the difference of Figs. 2.1 and 6.1; in Fig. 6.1 the states (wave functions) of the particles are shown as waves or solutions of the Schrödinger equation.

In this chapter we are concerned with the methods of counting of *arrangements of elements* taking into account the number of elements one can put into a state, their distinguishability or indistinguishability, and whether their number is conserved or not. These conditions lead to the three different types of statistics, Maxwell–Boltzmann, Fermi–Dirac and Bose–Einstein.

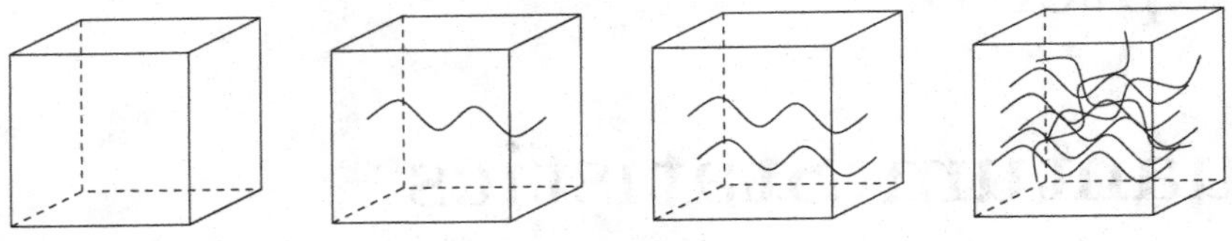

Particles in a box in quantum mechanics

no particle   1 particle   2 particles   N particles

Fig. 6.1 Particles in a box described by the Schrödinger equation.

6.2   *A priori* Weighting in Quantum Statistics

We consider first some simple cases exhibiting and explaining the concept of *degeneracy*. Consider an isolated system and suppose the level of energy ϵ_i has g_i degenerate states, *i.e.* it is g_i-fold degenerate. In other words, we have g_i different wave functions, *i.e.* solutions of the Schrödinger equation, with the same energy ϵ_i. Thus, for instance, for a hydrogen-like atom the spectrum (which is also illustrated in Fig. 6.2) is given by:

$$\epsilon_n = -\left(\frac{e^2}{\hbar c}\right)^2 \frac{mc^2}{2n^2}, \quad n = 1, 2, 3, \dots \infty, \qquad (6.3a)$$

where $n = n_r + l + 1$, $n_r = 0, 1, 2, \dots$, being the radial quantum number and $l = 0, 1, 2, \dots$, that of orbital angular momentum. The degeneracy of the associated quantum states is given by the expression

$$\sum_{l=0}^{n-1} (2l + 1) = n(n - 1) + n = n^2. \qquad (6.3b)$$

Thus[*] there are n^2 different weight functions giving the energy ϵ_n. Therefore $g_n = n^2$. Each of these degenerate levels is equally important. We therefore take the *a priori* probability introduced classically in Sec. 3.2 as proportional to the number of states concerned (this is really the fundamental postulate

---

[*]The degeneracy of hydrogen-like states is extensively discussed in H.J.W. Müller–Kirsten [47]. Or see A. Messiah [43], Vol. I, Sec. 11.1.4.

of equal *a priori* probabilities of an isolated system and will become clearer in the context of Sec. 11.2.2). The spectrum of these degenerate states of the hydrogen atom is illustrated in Fig. 6.2; for a consideration of the electronic partition function of the hydrogen atom see Example 6.5 — but note the problem associated with this. Since this is however a one-particle problem, we also recall from quantum mechanics the spectra of higher dimensional isotropic harmonic oscillators. The calculation of the spectra can be found in the treatise of Messiah [43] (there Sec. 12.3.1). Here we are only interested in the results in order to understand the aspect of degeneracy.

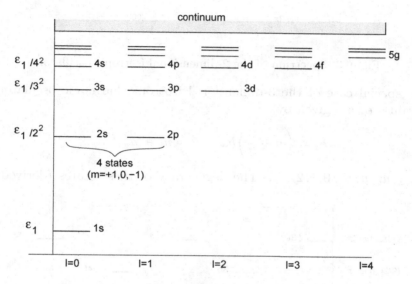

Fig. 6.2 The spectrum of the hydrogen atom.

Thus in the case of a p-dimensional isotropic harmonic oscillator with Hamiltonian (of p simple harmonic oscillators)

$$H = \sum_{i=1}^{p} \frac{1}{2m}(p_i^2 + m^2\omega^2 q_i^2) \tag{6.4}$$

the eigenvalues ϵ_N, $N = \sum_{i=1}^{p} n_i$, $n_1, n_2, n_3, \ldots, n_p = 0, 1, 2, \ldots$, are given by

$$\epsilon_N = \left[N + \frac{1}{2}p\right]\hbar\omega. \tag{6.5}$$

The spectrum of these eigenvalues and the degeneracy of the associated states for $p = 2$ is shown in Fig. 6.3.

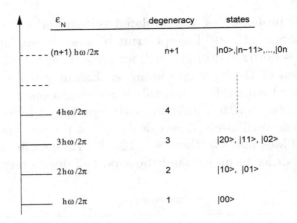

Fig. 6.3 Spectrum of the 2-dimensional isotropic oscillator.

In the special case of the 3-dimensional isotropic harmonic oscillator[†] the eigenvalues ϵ_n are given by

$$\epsilon_n = \left(n + \frac{3}{2}\right)\hbar\omega, \quad n = n_x + n_y + n_z, \tag{6.6}$$

with $n_x, n_y, n_z = 0, 1, 2, \ldots$. The degeneracy of these states (derived later

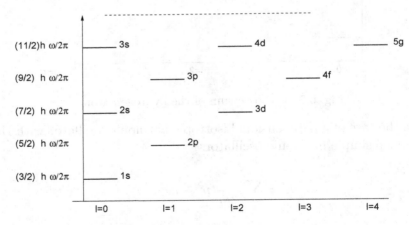

Fig. 6.4 Spectrum of the 3-dimensional isotropic oscillator.

in Example 12.1) is given by[‡]

$$\frac{1}{2}(n + 1)(n + 2). \tag{6.7}$$

[†] A. Messiah [43], Vol. I, Sec. 12.3.3.
[‡] See also A. Messiah [43], Vol. I, Sec. 12.3.3.

The spectrum of this case is shown in Fig. 6.4 and may be compared with that of the hydrogen atom. The result (6.7) will be required in Example 9.3.

6.2.1 Approximate calculation of number of states

We recall *Heisenberg's uncertainty principle*. If δq is the error or uncertainty in the spatial coordinate and δp correspondingly the error or uncertainty in the momentum coordinate in the same direction, this principle says

$$\delta q \delta p \geq h \tag{6.8}$$

for each direction. We distinguish between $\delta q, \delta p$ and some $\triangle q, \triangle p$, the latter being $\delta q, \delta p$ of some cell units in phase space, so that $\triangle q \triangle p / h$ is a certain number, as *e.g.* 4 in Fig. 6.5. In view of the uncertainty relation we cannot distinguish between these 4 entities, this means we cannot give them labels that permit us to distinguish one from the other. More generally, since the uncertainties $\delta q, \delta p$ can be anything between zero and infinity, it is not possible to distinguish between particles. In classical mechanics we could do that because there the uncertainties shrink to zero and we can pinpoint a particle exactly and give it a name or label. Therefore the number of *indistinguishable* states for each direction can be taken as $\triangle q \triangle p / h$.

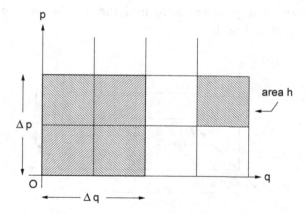

Fig. 6.5 Four states contained in $\triangle q \triangle p$.

Consider now, as an example, a *free* particle in 3-dimensional space. Since the particle is free it has no potential energy, and the number of states is

$$\frac{1}{h^3} \triangle q_x \triangle q_y \triangle q_z \times \triangle p_x \triangle p_y \triangle p_z.$$

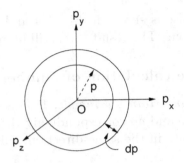

Fig. 6.6 The spherical momentum space element.

Since the potential energy is zero this can be integrated over ordinary space to give the volume V. Instead of the Cartesian momentum element $dp_x dp_y dp_z$ we take the volume of a spherical momentum element as depicted in Fig. 6.6. This element has volume $4\pi p^2 dp$. Hence the number of states in the total momentum range between p and $p + dp$ is

$$V \frac{4\pi p^2 dp}{h^3}. \tag{6.9}$$

6.2.2 Accurate calculation of number of states

We now consider an accurate evaluation of the number of states per unit momentum range. The problem is to find the number of *normal modes of vibration* in the box of Fig. 6.7.

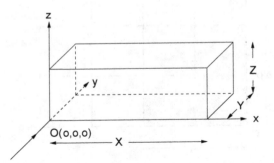

Fig. 6.7 The box-like volume element.

In 1924 de Broglie postulated: Associated with every particle there is a matter wave. If p is the momentum of the particle and λ the wavelength of the corresponding matter wave, then, so de Broglie,

$$p = \frac{h}{\lambda}.$$

If ψ is the amplitude of the matter wave, its Schrödinger wave equation is

$$\nabla^2\psi + \beta^2\psi = 0, \quad \text{where} \quad \beta = \frac{\omega}{c}. \tag{6.10}$$

Here ω is the circular frequency $= 2\pi\times$ ordinary frequency ν, and c is the phase velocity of the wave. Thus with $c = \nu\lambda$ we have

$$\beta = \frac{2\pi\nu}{c} = \frac{2\pi}{\lambda}.$$

For a Cartesian coordinate system Eq. (6.10) is

$$\frac{\partial^2\psi}{\partial x^2} + \frac{\partial^2\psi}{\partial y^2} + \frac{\partial^2\psi}{\partial z^2} + \beta^2\psi = 0. \tag{6.11}$$

As solution we take

$$\psi \propto \sin\left(\frac{l\pi x}{X}\right) \sin\left(\frac{m\pi y}{Y}\right) \sin\left(\frac{n\pi z}{Z}\right), \tag{6.12}$$

or with cosines. The expression with sines is more suitable for the analysis to follow. Hence by substitution into Eq. (6.11) we obtain (*cf.* Fig. 6.8)

$$\frac{l^2}{X^2} + \frac{m^2}{Y^2} + \frac{n^2}{Z^2} = \frac{\beta^2}{\pi^2}, \tag{6.13}$$

where l, m, n are positive integers unequal to zero, to prevent duplication.

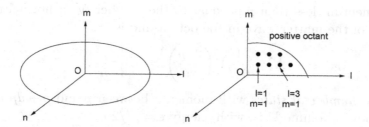

Fig. 6.8 The ellipsoid and its positive octant.

Hence if l, m, n are integers, the amplitude is zero on all walls of the box (where $x = X, y = Y, z = Z$). The vanishing of the Schrödinger wave function on all walls of the box is the quantum mechanical boundary condition which confines the particle to the inside of the box. Note that here the box is defined by simple 3-dimensional Cartesian coordinates. In the case of a spherical volume the solution (6.12) would involve a spherical Bessel function, and the eigenvalues $\propto \beta^2$ would be determined by the zeros of this function (see Examples 6.13 and 7.9). Now consider a range of β corresponding to a range of λ. We want to find the number of solutions for which the inequality

$$\frac{l^2}{X^2} + \frac{m^2}{Y^2} + \frac{n^2}{Z^2} < \frac{\beta^2}{\pi^2} \tag{6.14}$$

is satisfied. We have (note that $\beta X, \beta Y, \beta Z$ are dimensionless)

$$\frac{l^2}{(\beta X/\pi)^2} + \frac{m^2}{(\beta Y/\pi)^2} + \frac{n^2}{(\beta Z/\pi)^2} < 1. \tag{6.15}$$

This is the equation for the interior of an ellipsoid, *i.e.* for the region within an ellipsoid with semi-axes of lengths $\beta X/\pi, \beta Y/\pi, \beta Z/\pi$. We deal only with the positive octant, *i.e.* with positive l, m, n, as indicated in Fig. 6.8. Since the spacing between the l's is unity, and similarly that for the m's and n's, the size of a unit cell is 1. Hence the number of (l, m, n) values[§] is approximately equal to the volume of the octant. This is valid only if the number of points within the octant is very large — the case we consider here. The volume of the (non-spatial) octant $V_{1/8}$, with $V = 8V_{1/8}$, is[¶]

$$V_{1/8} = \frac{1}{8}\frac{4}{3}\pi\left(\frac{\beta X}{\pi}\right)\left(\frac{\beta Y}{\pi}\right)\left(\frac{\beta Z}{\pi}\right)$$

$$= XYZ\frac{1}{6}\beta^3\frac{1}{\pi^2} = V\frac{1}{6}\frac{\beta^3}{\pi^2}$$

$$= \text{number of states.} \tag{6.16}$$

Since $p = h/\lambda = \beta h/2\pi$, we have $\beta = 2\pi p/h$. Hence the number of states with momentum less than p is equal to the number $\mathcal{N}$ of normal modes of vibration of the matter wave in the octant and is

$$\mathcal{N} = \frac{1}{6\pi^2}V\left(\frac{2\pi p}{h}\right)^3 = V\frac{4}{3}\pi\frac{p^3}{h^3}.$$

Hence the *number of states* with momenta between p and $p + dp$ in a rectangular box of volume V is, with energy $\epsilon = p^2/2m$,

$$d\mathcal{N} = V\frac{4\pi p^2 dp}{h^3} = \frac{2\pi V(2m)^{3/2}}{h^3}\epsilon^{1/2}d\epsilon,$$

and this is equal to the value of g corresponding to a certain momentum range. The result is seen to agree with that in Eq. (6.9), and with the result

---

[§]This number expressed in terms of the wave number β is called the number of *normal modes*; expressed in terms of momentum it is called the number of *wave-mechanical states*.

[¶]Recall that the area of a circle of radius a is πa^2 and correspondingly the area of an ellipse of semi-major and minor axes a and b is πab. Similarly the volume of a sphere of radius a is $4\pi a^3/3$ and that of an ellipsoid of semi-axes a, b, c is $4\pi abc/3$.

(3.9c), the corresponding *a priori* probability. Thus this number of states, the degree of degeneracy, is given by the *a priori* probability as stipulated earlier. For other treatments of the counting of quantum states see *e.g.* K. Huang [35].

We summarize the above findings in the following statements:

1. We deduce from *Heisenberg's uncertainty principle* that in the approximation in which energy levels are taken to form a continuum, we have

$$\sum_i g_i f(\epsilon_i) = \int \int \cdots \int f(\epsilon) \frac{(dq_1 dq_2 \cdots dq_n)(dp_1 dp_2 \cdots dp_n)}{h^n}, \quad (6.17)$$

    where ϵ_i, $i = 1, 2, 3, \ldots$, are the possible energy levels of the elements comprising the system, n is the number of dimensions of the system, and gf is a distribution function, and g the *a priori* probability.[|]

2. As an illustration we deduce from Eq. (6.17) that for a free atom in a 3-dimensional volume V,

$$\sum_i g_i f(\epsilon_i) = \frac{4\pi V}{h^3} \int f\left(\frac{p^2}{2m}\right) p^2 dp = \frac{2\pi V (2m)^{3/2}}{h^3} \int \epsilon^{1/2} f(\epsilon) d\epsilon.$$

$$(6.18)$$

    This result is confirmed by finding from Schrödinger's equation the energy levels of an atom in a rectangular box and making an approximate evaluation of the number of levels per unit energy range, as we did in the preceding considerations.

6.2.3 Examples

Example 6.1: Number of quantum states in one dimension

Calculate the approximate number of normal modes with wave numbers between β and $\beta + d\beta$ associated with the one-dimensional Schrödinger equation

$$\frac{\partial^2 \psi(x)}{\partial x^2} + \beta^2 \psi(x) = 0, \quad \beta = \frac{2\pi}{\lambda}, \quad (6.19)$$

in the domain $0 < x < L$. Show that the result could have been obtained by considering the number of half-wavelengths which could be fitted into the length L. Why would the corresponding number of wave-mechanical states be $L dp_x / h$, and not $2L dp_x / h$, as might appear at first sight?

Solution: We have

$$\frac{\partial^2 \psi(x)}{\partial x^2} + \beta^2 \psi(x) = 0, \quad (6.20)$$

[|] Recall that in mathematical terms, probability is a *measure* defined for each event, like length, area or volume of a region in one, two or three dimensions. For a non-technical discussion of how the concept of probability is arrived at see A. Ben–Naim [4], pp.25-30.

and we choose $\psi(x) \propto \sin(l\pi x/L)$ with l an integer so that $\psi(0) = 0$ and $\psi(L) = 0$. Then

$$\frac{\partial^2 \psi}{\partial x^2} \propto -\frac{l^2 \pi^2}{L^2}\psi,$$

and therefore

$$\beta^2 = \frac{l^2 \pi^2}{L^2}, \qquad \beta = (\overset{+}{-})\frac{l\pi}{L}. \tag{6.21}$$

Therefore the integer l which is equal to the number of normal modes in the domain $0 < x < L$ is given by

$$l = \frac{L\beta}{\pi}. \tag{6.22}$$

Then

$$dl = \frac{L d\beta}{\pi}, \tag{6.23}$$

which is the requested number of normal modes. The number of half wavelengths, $\lambda/2$, which could be fitted into L is (since $\beta = 2\pi/\lambda$)

$$\frac{L}{\lambda/2} = \frac{2L}{\lambda} = \frac{L\beta}{\pi}. \tag{6.24}$$

Hence as before we obtain $L d\beta/\pi$. Now consider the wave mechanical states with momentum p_x. Here

$$p_x = \frac{h}{\lambda} = \frac{\beta h}{2\pi}, \qquad \beta = \frac{2\pi p_x}{h}. \tag{6.25}$$

Therefore at first sight:

$$\frac{L d\beta}{\pi} = \frac{L 2\pi dp_x}{h\pi} = \frac{2L dp_x}{h}. \tag{6.26}$$

Here we have to divide by 2 in order to get the number of wave-mechanical states. The reason is that p_x in Eq. (6.25) can be positive or negative but β is to be positive. This requires p_x or L to be divided by 2 here. This is the 1-dimensional reduction of the 2- and 3-dimensional cases of Example 6.2 and Sec. 6.2.2 respectively.

Example 6.2: Number of quantum states in two dimensions

By considering the solutions of the two-dimensional Schrödinger equation

$$\frac{\partial^2 \psi(x,y)}{\partial x^2} + \frac{\partial^2 \psi(x,y)}{\partial y^2} + \beta^2 \psi(x,y) = 0 \tag{6.27}$$

for the rectangular region $0 < x < X, 0 < y < Y$, show that the number of normal modes in the wave number domain between β and $\beta + d\beta$ is approximately $A\beta d\beta/2\pi$, where A is the area XY. Show that the corresponding number of wave-mechanical states is $2\pi A p dp/h^2$.

Solution: In this case we have

$$\frac{\partial^2 \psi(x,y)}{\partial x^2} + \frac{\partial^2 \psi(x,y)}{\partial y^2} + \beta^2 \psi(x,y) = 0. \tag{6.28}$$

We choose

$$\psi(x,y) \propto \sin\left(\frac{l\pi x}{X}\right) \sin\left(\frac{m\pi y}{Y}\right). \tag{6.29}$$

Inserting this into the wave equation we obtain

$$\frac{l^2}{X^2} + \frac{m^2}{Y^2} = \frac{\beta^2}{\pi^2}. \tag{6.30}$$

The domain within the ellipse shown in Fig. 6.8 is therefore given by the inequality

$$\frac{l^2}{(\beta X/\pi)^2} + \frac{m^2}{(\beta Y/\pi)^2} \leq 1. \tag{6.31}$$

Since the spacing between the l values is unity and similarly that of the m's, the size of a unit cell is 1. Hence the number of (l, m)-values is approximately equal to the area of a quadrant of the ellipse $(l, m \geq 0)$. The area of the quadrant is

$$\frac{1}{4}\pi\frac{\beta^2 XY}{\pi^2} = \frac{1}{4}\frac{\beta^2 A}{\pi}, \quad A = XY. \tag{6.32}$$

This is the number of states. Now

$$p = \frac{h}{\lambda} = \frac{\beta h}{2\pi}, \quad \therefore \beta = \frac{2\pi p}{h}. \tag{6.33}$$

It follows that the number of states with β between β and $\beta + d\beta$ is

$$\frac{d}{d\beta}\left[\frac{1}{4}\frac{\beta^2 A}{\pi}\right]d\beta = \frac{1}{2}\frac{A\beta d\beta}{\pi} = \frac{1}{2}A\frac{2\pi p}{h\pi}\frac{2\pi dp}{h} = \frac{2\pi A p dp}{h^2}. \tag{6.34}$$

This result could also have been obtained by multiplying $\delta x \delta y/h^2 = A/h^2$ by $\delta p_x \delta p_y = 2\pi p dp$ (this is the 2-dimensional version of the 3-dimensional case of Sec. 6.2.1).

Example 6.3: Density of states of the harmonic oscillator

Assuming that the number of quantum states in a range $\delta q \delta p$ for each direction of motion is given, per element, by a factor $\delta q \delta p/h$, show that the number of states in the energy range $d\epsilon$ is for a simple harmonic oscillator of natural frequency ν given by $d\epsilon/h\nu$ as deduced from Eq. (3.12). From elementary quantum mechanics it is known that the energy levels of the simple harmonic oscillator are given by $\epsilon_n = h\nu(n + 1/2)$, and are non-degenerate. Calculate approximately the number of states in range $d\epsilon$ by evaluating $\partial n/\partial \epsilon = 1/(\partial \epsilon/\partial n)$.

Solution: As given, we have $g = d\epsilon/\nu = dqdp$. Hence

$$\frac{dqdp}{h} = \frac{d\epsilon}{h\nu}.$$

The number of states in the range $d\epsilon$ is $d\epsilon/h\nu$. Since $\epsilon_n = h\nu(n + 1/2)$, we have $\partial \epsilon_n/\partial n = h\nu$, so that

$$\frac{1}{\partial \epsilon/\partial n} = \frac{1}{h\nu} = \frac{\partial n}{\partial \epsilon}, \quad \frac{d\epsilon}{h\nu} = dn. \tag{6.35}$$

Example 6.4: Number of states of a diatomic molecule in the range $d\epsilon$

Assuming that the number of quantum states in a range $\delta q \delta p$ for each direction of motion is given, per element, by a factor $\delta q \delta p/h$, show that the number of states in the energy range $d\epsilon$ is $8\pi^2 I d\epsilon/h^2$ for a rotating diatomic molecule (*cf.* Example 3.6). It is known from wave mechanics that the energy levels of a rotating diatomic molecule are $\epsilon_J = J(J + 1)h^2/8\pi^2 I$, each such level being $(2J + 1)$-fold degenerate. Calculate approximately the number of states in range $d\epsilon$ by evaluating $\partial J/\partial \epsilon = 1/(\partial \epsilon/\partial J)$ and multiplying by the corresponding degeneracy.

Solution: From Example 3.6, Eq. (3.25), we know that in the case of the diatomic molecule of moment of inertia I the *a priori* weighting g is given by $g \propto 8\pi^2 I d\epsilon$. In terms of polar and azimuthal angles θ and ϕ we have $g \propto dq_\theta dq_\phi dp_\theta dp_\phi$. Here the first two factors are of no importance in view of symmetry conditions. Therefore we have

$$\frac{g}{h^2} = \frac{8\pi^2 I d\epsilon}{h^2}. \tag{6.36}$$

Since
$$\epsilon_J = \frac{J(J+1)h^2}{8\pi^2 I}, \qquad \frac{\partial \epsilon_J}{\partial J} = \frac{(2J+1)h^2}{8\pi^2 I}, \qquad \frac{\partial J}{\partial \epsilon_J} = \frac{8\pi^2 I}{(2J+1)h^2}, \qquad (6.37)$$

we obtain
$$(2J+1)\frac{\partial J}{\partial \epsilon_J} = \frac{8\pi^2 I}{h^2}. \qquad (6.38)$$

Therefore the number of states in the range $d\epsilon$ is approximately $(2J+1)dJ = 8\pi^2 I d\epsilon/h^2 = g d\epsilon/h^2$.

Example 6.5: The partition function for the hydrogen atom

In Sec. 6.2 the hydrogen atom was referred to and its discrete energy eigenvalues were given and the degeneracy n^2 was cited. Establish the electronic partition function (defined in Eq. (5.13)) and consider its sum.

Solution: The energy of the n-th energy level of the hydrogen atom, as derived in books on quantum mechanics, is $\epsilon_n = -R_\infty/n^2$, where R_∞ is the constant, known as *Rydberg constant*, given in Eq. (6.3a). The partition function Z was defined in Eq. (5.13). Measuring the excited energies from the ground state $n = 1$ upwards, the *electronic partition function* per atom is

$$Z_{el} = \sum_n n^2 \exp\left[-\frac{R_\infty}{kT}\left(1 - \frac{1}{n^2}\right)\right]. \qquad (6.39)$$

Here the nondegenerate ground state energy has been subtracted out. We observe that owing to the degeneracy $g = n^2$, the sum does not converge, implying that the sum is infinite. This cannot be correct, and may be another one of the divergence problems associated with the Coulomb potential and solved in field theory by vacuum polarization.**

6.3   The Allowed Number of Elements in Quantum States

6.3.1   One element

Consider one particle (element) 'a'. This particle will be represented by a wave function. The wave function of the matter wave has an amplitude $\psi(a)$. The intensity is proportional to the square of the amplitude (assumed real as in the previous section). The probability of finding the element per unit volume is equal to

$$\psi_n^2(a),$$

or, more generally, as in the case when ψ is complex, $|\psi_n(a)|^2$. The quantum number n designates the energy level in which the element is situated.

6.3.2   Two non-interacting elements

We consider non-interacting elements, like *e.g.* the non-interacting atoms or molecules of a perfect gas. Let the particles be 'a' and 'b'. At first sight one

---

**For a simple discussion see R.P.H. Gasser and W.G. Richards [28], p.45.

would think the wave function of the pair is

$$\Psi_1 = \psi_n(a)\psi_m(b),$$

where Ψ_1 is the total wave function for the system consisting of the two non-interacting elements. However, Ψ_1 is only one solution. There is another solution

$$\Psi_2 = \psi_m(a)\psi_n(b).$$

The energy of the 2-particle state is the same in both cases, but the (identical) elements are interchanged. This solution is therefore just as good as the first (m and n being quantum numbers).

What linear combination forms the actual wave function? The condition that the interchange of 'a' and 'b' labels is unobservable is that the observed probability

$$\Psi^2(a, b) \quad \text{or} \quad |\Psi(a, b)|^2$$

is independent of the interchange of 'a' and 'b'. Hence

$$\Psi^2(a, b) = \Psi^2(b, a), \tag{6.40}$$

and therefore

$$\Psi(a, b) = \pm\Psi(b, a). \tag{6.41}$$

Hence the conditions are satisfied only by

$$\Psi(a, b) \to \Psi_\pm(a, b) = \Psi_1 \pm \Psi_2 = \psi_n(a)\psi_m(b) \pm \psi_m(a)\psi_n(b). \tag{6.42}$$

Here the + sign gives a *symmetrical total wave function*, and the - sign an *antisymmetrical total wave function*. The transition probability between wave functions of different symmetry is zero, because

$$\int dV \Psi_+(a, b)\Psi_-(a, b) = \int dV \psi_n^2(a)\psi_m^2(b) - \int dV \psi_m^2(a)\psi_n^2(b) = 0.$$

For practically all basic real elementary particles, *e.g.* electrons, protons *etc.* (but also quarks), the spatial wave functions are antisymmetrical and give rise to *Fermi–Dirac statistics*. Non-real elements, such as photons and phonons, but also mesons (*e.g.* pions) obey the spatially symmetrical *Bose–Einstein statistics*. For Bose–Einstein statistics it is possible to have $m = n$, so that

$$\Psi \propto \psi_n(a)\psi_n(b)$$

for any number of elements in the same state. In the case of Fermi–Dirac statistics with the antisymmetrical total wave function, putting $n = m$ implies

$$\Psi \propto \psi_n(a)\psi_n(b) - \psi_n(a)\psi_n(b) = 0.$$

Thus this number state corresponds to that of the *Pauli exclusion principle*. This says, it is impossible to put more than one element into a given quantum state.

6.3.3 More than two elements stuck together

In the case of *polyatomic molecules* which contain identical atoms only such states are realizable in nature for which the total wave functions are (a) symmetric if the mass number (*i.e.* number of elementary particles) of the atoms is even, or (b) antisymmetric if this mass number is odd. An even number of particles stuck together obeys Bose–Einstein statistics in spite of the fact that each of them separately obeys Fermi–Dirac statistics. This follows from the fact that an even number of changes of sign in the total wave function produces no change in sign. For example, Helium with atomic weight 4 has four elementary particles contained in its nucleus, *i.e.* an even number of elementary particles in the nucleus. Hence He obeys Bose–Einstein statistics, although the elementary particles of which it is composed obey Fermi–Dirac statistics. Hence Bose–Einstein statistics can also apply to real atoms.

6.4 Counting of Number of Arrangements

We now consider the counting of the *number of arrangements*[††] of a number of elements in quantum statistics, *e.g.* N. We assume

1. The elements are indistinguishable from one another, *i.e.* they have no labels.

2. In the case of Bose–Einstein statistics we can have any number of elements in the same state. In the case of Fermi–Dirac statistics we have either 0 or 1 element in each state.

6.4.1 Fermi–Dirac statistics

We begin with Fermi–Dirac statistics. *Temporarily* attach labels to the elements. Let us deal with one energy level of a one-particle system of energy ϵ_i by itself. Let the degeneracy of this energy level ϵ_i be g_i, *i.e.* g_i states all have the same energy ϵ_i, as indicated in Fig. 6.9. Suppose there are n_i elements to be distributed amongst the g_i states (temporarily with labels attached).

[††]For other discussions of the relations W_{FD}, W_{BE} which we establish here, see *e.g.* F. Mandl [41], p.276.

Stick the first element in any one of the g_i states (since we are considering Fermi–Dirac statistics this one element is the only one we can put into this state), stick the second element in any one of the other $(g_i - 1)$ states (again only one element is admissable),
....

Therefore the number of arrangements is

$$
\begin{aligned}
&= g_i(g_i - 1)(g_i - 2) \cdots \quad \text{to } n_i \text{ factors} \\
&= g_i(g_i - 1)(g_i - 2) \cdots (g_i - n_i + 1) \\
&= \frac{g_i!}{(g_i - n_i)!}.
\end{aligned}
$$

To remove the labels attached to these elements we divide this number of arrangements of n_i particles with labels by the number of arrangements of n_i elements amongst themselves, *i.e.* $n_i!$.

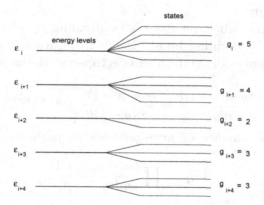

Fig. 6.9 Energy levels with energy ϵ_i and degeneracy g_i.

Therefore the number of arrangements required is:

$$
\frac{g_i!}{n_i!(g_i - n_i)!} = {}^{g_i}C_{n_i} = \binom{g_i}{n_i}. \tag{6.43}
$$

Therefore, taking into account *all* the energy levels of the total number of indistinguishable elements $N = \sum_i n_i$, the number of arrangements becomes

$$
\prod_i \frac{g_i!}{n_i!(g_i - n_i)!} = W_{FD}. \tag{6.44}
$$

This expression is referred to as the *number of arrangements of particles in Fermi–Dirac statistics*.

6.4.2 Bose–Einstein statistics

Next we consider Bose–Einstein statistics. *Temporarily* we attach labels to
both states and elements. A typical arrangement of elements amongst states
is (omitting indices i):

$$g^{(4)}n^{(1)}n^{(27)}g^{(2)}n^{(2)}n^{(4)}n^{(5)}g^{(13)}g^{(3)}.$$

Here $g^{(4)}$ stands for "state labelled '4' ", and (note the order of g's and n's)
$n^{(1)}, n^{(27)}$ stand for "elements numbers (1) and (27) put into state $g^{(4)}$".
In the given arrangement there are no elements in state $g^{(13)}$! Hence in
describing our arrangements we must always start with a state, *i.e.* with
a g. The remaining $(g + n - 1)$ quantities can be arranged in any way.
Therefore:

$$\text{number of arrangements } = g(g + n - 1)!, \tag{6.45a}$$

where the factor g on the right gives the number of ways of choosing a g to
be the first quantity.

   Now remove the labels on the n's by dividing by $n!$. Also remove the
labels on the g's by dividing by $g!$. Hence if both states and elements are
unlabelled, the number of arrangements is (observe the ordering, with g first,
not n)

$$\frac{g(g + n - 1)!}{n!g!} = \frac{(g + n - 1)!}{n!(g - 1)!} = {}^{g+n-1}C_n, \tag{6.45b}$$

and therefore the *number of arrangements of particles in Bose–Einstein
statistics* is

$$W_{BE} = \prod_i \frac{(g_i + n_i - 1)!}{n_i!(g_i - 1)!}. \tag{6.46}$$

Both this and the Fermi–Dirac number of arrangements (6.44) should now
be compared with the Maxwell–Boltzmann number of arrangements given
by Eq. (4.2), *i.e.*

$$W_{MB} = N! \prod_i \frac{g_i^{n_i}}{n_i!}. \tag{6.47}$$

It is particularly instructive to consider the comparison at high temperatures.
This is done in the next section.

6.5 Quantum Statistics at High Temperatures

We want to consider the relationships between the W's at high temperatures.
At high temperatures plenty of energy is available and the high energy levels
are very often occupied (ϵ is large). Therefore over all states the dilution is

very great. Plenty of states are effectively available. Therefore the probability of any state being occupied is small. Therefore the maximum occupation of energy level ϵ_i is g_i, *i.e.* $n_i = g_i$. Therefore at high temperatures n_i is normally less than this maximum, *i.e.*:

$$n_i \ll g_i. \tag{6.48}$$

Consider W_{FD}. In this case

$$\frac{g!}{n!(g-n)!} = \frac{g(g-1)(g-2)\cdots \text{ to } n \text{ factors}}{n!} \sim \frac{g^n}{n!}, \tag{6.49}$$

where $\sim$ means 'of order of magnitude'. Thus at high temperatures

$$W_{FD} \rightarrow \prod_i \frac{g_i^{n_i}}{n_i!}. \tag{6.50}$$

However, at all temperatures:

$$W_{MB} = N! \prod_i \frac{g_i^{n_i}}{n_i!}. \tag{6.51}$$

Similarly at high temperatures:

$$W_{BE} \rightarrow \prod_i \frac{g_i^{n_i}}{n_i!} \tag{6.52}$$

Hence *the classical Maxwell–Boltzmann result is wrong*, even at high temperatures it is wrong by the factor $N!$. In classical statistics the elements are erroneously regarded as distinguishable which, of course, is wrong according to the Heisenberg uncertainty principle (*cf.* the extra factor $N!$ in W_{MB}).[‡‡]

We now *consider the removal* of the factor $N!$ from previous classical results (*e.g.* W_{MB}). Then there is no difference in occupation numbers, *i.e.* (*cf.* Eqs. (4.10), (4.14))

$$n_i = g_i e^{-\epsilon_i/kT} \qquad \text{for nonconserved elements,}$$

$$n_i = \frac{N g_i e^{-\epsilon_i/kT}}{\sum_i g_i e^{-\epsilon_i/kT}} \qquad \text{for conserved elements.} \tag{6.53}$$

Here we assume there are no external fields and no collisions between particles. Therefore $f_i = n_i/g_i$ is time-independent, and $df_i/dt = 0$. In other

---

[‡‡]This does not mean that Maxwell–Boltzmann statistics is always wrong — it is wrong only in the case of indistinguishable elements. For a case in which Maxwell–Boltzmann statistics is appropriate see Example 9.2.

cases — see Boltzmann equation — still $df_i/dt = 0$, but $f = f_i(t, velocity)$. However, the removal of the factor $N!$ does affect the entropy S. We consider this now. Since $S_{MB} = k \ln W_{MB}$, we must subtract $k \ln N!$ from S_{MB}. We had (*cf.* Eqs. (5.11), (5.9)) for the Maxwell–Boltzmann entropy

$$S_{MB} = \frac{E}{T} + Nk \ln \sum_i g_i e^{-\epsilon_i/kT}. \tag{6.54}$$

Therefore the high temperature quantum statistics entropy S is given by:

$$S \simeq S_{MB} - k \ln N! \simeq S_{MB} - kN(\ln N - 1), \tag{6.55}$$

where we used Stirling's theorem $\ln N! \simeq N(\ln N - 1)$ for N large. Therefore from Eqs. (5.11), (5.9):

$$S = \frac{E}{T} + Nk \left[1 + \ln \left(\frac{\sum_i g_i e^{-\epsilon_i/kT}}{N} \right) \right]. \tag{6.56}$$

From this we obtain the *high temperature* free energy F as

$$F \stackrel{(5.10)}{=} E - TS = F_{MB} + kT \ln N! = -NkT \left[1 + \ln \left(\frac{\sum_i g_i e^{-\epsilon_i/kT}}{N} \right) \right]. \tag{6.57}$$

This is the free energy F with the factor $N!$ removed from W_{MB}. We can now consider applications.

6.6   Applications

We apply the above results to *free particles*, *i.e.* here to a monatomic gas, in a volume V. In this case

$$g = V \frac{4\pi p^2 dp}{h^3}, \tag{6.58}$$

and therefore, using one of the results (2.20),

$$\sum_i g_i e^{-\epsilon_i/kT} = \frac{4\pi V}{h^3} \int_0^\infty e^{-p^2/2mkT} p^2 dp = \frac{V}{h^3} (2\pi mkT)^{3/2}. \tag{6.59}$$

By a corresponding direct evaluation of

$$E = \sum_i n_i \epsilon_i,$$

or by equipartition of energy, we obtain

$$E = \frac{3}{2}NkT. \tag{6.60}$$

The entropy S of a vapour is therefore given by

$$
\begin{aligned}
S \overset{(6.56)}{=}\; & \frac{E}{T} + Nk\left[1 + \ln\left(\frac{\sum_i g_i e^{-\epsilon_i/kT}}{N}\right)\right] \\
\overset{(6.60)}{=}\; & Nk\left[1 + \frac{3}{2} + \ln\left(\frac{\sum_i g_i e^{-\epsilon_i/kT}}{N}\right)\right] \\
\overset{(6.59)}{=}\; & Nk\left[\frac{5}{2} + \ln\left\{\frac{V}{Nh^3}(2\pi mkT)^{3/2}\right\}\right] \\
=\; & Nk\left[\frac{3}{2}\ln T + \ln\frac{V}{N} + \frac{5}{2} + \frac{3}{2}\ln\left(\frac{2\pi mk}{h^3}\right)\right]. \tag{6.61}
\end{aligned}
$$

This formula[§§] is known as the *Sackur–Tetrode formula*.[¶¶] From this formula, valid at temperatures low enough to permit neglecting the entropy of the condensed states, we can derive below the vapour pressure. We combine this with the expression for the *latent heat* L_{vapour} of a vapour. This latent heat is as in Eq. (5.52) defined as the quantity:

$$L_{\text{vapour}} = \frac{1}{T}(S_{\text{vapour}} - S_{\text{liquid}}), \tag{6.62}$$

where S_{liquid} is very small and almost constant. With (*cf.* Eq. (5.17))

$$P_{\text{vapour}} = T\left(\frac{\partial S}{\partial V}\right)_E = \left(\frac{N}{V}\right)kT \tag{6.63}$$

the vapour pressure is obtained in terms of L_{vapour} and T.

6.7 Summary

We summarize our results for the distributions:

$$W_{BE} = \prod_i \frac{(n_i + g_i - 1)!}{(g_i - 1)!n_i!}, \tag{6.64}$$

[§§]Observe that the term $\ln V$ in the classical (Maxwell–Boltzmann) expression (5.47) (there per atom) is here in the quantum statistical result $\ln(V/N)$.

[¶¶]O. Sackur [64], H. Tetrode [76]. For further discussion see R.P.H. Gasser and W.G. Richards [28], pp.38-40, E. Schrödinger [65], p.57.

$$W_{FD} = \prod_i \frac{g_i!}{n_i!(g_i - n_i)!}. \tag{6.65}$$

We observe that in each case: The sum of terms in the numerator = the sum of terms in the denominator.

The results of high temperature quantum statistics are good enough for dilute systems (a system is dilute when its elements are at large distances away from one another), *i.e.* for perfect gases. But this does not·apply to solids and liquids. Hence use high temperature results for all gases.

6.8   Applications and Examples

Example 6.6: Estimating errors in high temperature approximations
By employing the methods of Examples 4.1, 4.2, estimate the respective errors incurred in approximating

$$W_{FD} = \prod_i \frac{g_i!}{(g_i - n_i)!n_i!} \quad \text{and} \quad W_{BE} = \prod_i \frac{(n_i + g_i - 1)!}{(g_i - 1)!n_i!} \tag{6.66}$$

by $\tilde{W} = \prod_i g_i^{n_i}/n_i!$.

Solution: Consider the *Fermi–Dirac distribution*. As argued before Eq. (6.48) high temperatures imply $g_i \gg n_i$. In this case we have, with the help of the Stirling formula $n! \simeq \sqrt{2\pi} n^{n+1/2} e^{-n}$,

$$\frac{g_i!}{(g_i - n_i)!n_i!} \simeq \frac{g_i^{g_i+1/2} e^{-g_i}}{g_i^{g_i-n_i+1/2} e^{-g_i+n_i}} \frac{1}{n_i!} \simeq \frac{g_i^{n_i}}{n_i! e^{n_i}} \simeq \frac{g_i^{n_i}}{n_i!}. \tag{6.67}$$

The error incurred by this approximation is $\triangle = W_{FD} - \tilde{W}$, *i.e.*

$$\triangle = \prod_i \frac{g_i!}{(g_i - n_i)!n_i!} - \prod_i \frac{g_i^{n_i}}{n_i!}$$

$$= \prod_i \frac{g_i^{g_i+1/2} e^{-g_i}}{\sqrt{2\pi}(g_i - n_i)^{g_i-n_i+1/2} e^{-g_i+n_i} n_i^{n_i+1/2} e^{-n_i}} - \prod_i \frac{g_i^{n_i}}{n_i!}$$

$$\simeq \prod_i \frac{g_i^{g_i+1/2}}{\sqrt{2\pi} g_i^{g_i-n_i+1/2} (1 - n_i/g_i)^{g_i-n_i+1/2} n_i^{n_i+1/2}} - \prod_i \frac{g_i^{n_i}}{n_i!}$$

$$\simeq \prod_i \frac{g_i^{n_i}}{\sqrt{2\pi} n_i^{n_i+1/2}} \left(1 - \frac{n_i}{g_i}\right)^{-g_i+n_i-1/2} - \prod_i \frac{g_i^{n_i}}{n_i!}$$

$$\simeq \prod_i \frac{g_i^{n_i}}{\sqrt{2\pi} n_i^{n_i+1/2}} \left(1 - \frac{n_i}{g_i}\right)^{-g_i+n_i-1/2} - \prod_i \frac{g_i^{n_i}}{\sqrt{2\pi} n_i^{n_i+1/2}} e^{n_i}. \tag{6.68}$$

In this expression

$$\left(1 - \frac{n_i}{g_i}\right)^{-g_i} \left\{ \left(1 - \frac{n_i}{g_i}\right)^{n_i-1/2} \right\} \overset{g_i \to \infty}{\simeq} \frac{1}{e^{-n_i}} \left\{ 1 - \frac{n_i(n_i - 1/2)}{g_i} \right\}, \tag{6.69}$$

where we used a formula of the exponential function given in the literature.*** Hence

$$\Delta \simeq \left(\prod_i \frac{g_i^{n_i}}{\sqrt{2\pi} n_i^{n_i+1/2}} e^{n_i} \right) \prod_i \left\{ 1 - \frac{n_i(n_i - 1/2)}{g_i} \right\} - \left(\prod_i \frac{g_i^{n_i}}{\sqrt{2\pi} n_i^{n_i+1/2}} e^{n_i} \right)$$

$$\simeq \left(\prod_i \frac{g_i^{n_i} e^{n_i}}{\sqrt{2\pi} n_i^{n_i+1/2}} \right) \left(- \sum_i \frac{n_i(n_i - 1/2)}{g_i} \right) = - \left(\sum_i \frac{n_i(n_i - 1/2)}{g_i} \right) \tilde{W}. \qquad (6.70)$$

Thus the error is of the order of $\tilde{W}/(\text{degeneracies } g_i)$, where $g_i \gg n_i$ for all i.

In the case of *Bose–Einstein statistics* we arrive at a similar result. We have with the help of Stirling's formula in the form given above:

$$\frac{(n_i + g_i - 1)!}{(g_i - 1)! n_i!} \simeq \frac{(n_i + g_i - 1)^{n_i + g_i - 1/2} e^{-n_i - g_i + 1}}{(g_i - 1)^{g_i - 1/2} e^{-g_i + 1} n_i!} \underset{g_i \gg n_i}{\simeq} \frac{g_i^{n_i}}{n_i! e^{n_i}} \simeq \frac{g_i^{n_i}}{n_i!}. \qquad (6.71)$$

The error incurred by this approximation is

$$\Delta = \prod_i \frac{(n_i + g_i - 1)!}{(g_i - 1)! n_i!} - \prod_i \frac{g_i^{n_i}}{n_i!} = \prod_i \frac{(n_i + g_i - 1)^{n_i + g_i - 1/2} e^{-n_i - g_i + 1}}{\sqrt{2\pi}(g_i - 1)^{g_i - 1/2} e^{-g_i + 1} n_i^{n_i + 1/2} e^{-n_i}} - \prod_i \frac{g_i^{n_i}}{n_i!}$$

$$= \prod_i \frac{g_i^{n_i + g_i - 1/2}[1 + (n_i - 1)/g_i]^{n_i + g_i - 1/2}}{\sqrt{2\pi} g_i^{g_i - 1/2} n_i^{n_i + 1/2}[1 - 1/g_i]^{g_i - 1/2}} - \prod_i \frac{g_i^{n_i}}{n_i!}$$

$$= \prod_i \frac{g_i^{n_i}[1 + (n_i - 1)/g_i]^{n_i + g_i - 1/2}}{\sqrt{2\pi} n_i^{n_i + 1/2}[1 - 1/g_i]^{g_i - 1/2}} - \prod_i \frac{g_i^{n_i}}{\sqrt{2\pi} n_i^{n_i + 1/2} e^{-n_i}}$$

$$\simeq \prod_i \frac{g_i^{n_i}}{\sqrt{2\pi} n_i^{n_i + 1/2}} \left(1 + \frac{n_i - 1}{g_i} \right)^{n_i + g_i - 1/2} \left(1 - \frac{1}{g_i} \right)^{1/2 - g_i} - \prod_i \frac{g_i^{n_i}}{\sqrt{2\pi} n_i^{n_i + 1/2}} e^{n_i}$$

$$\simeq \prod_i \frac{g_i^{n_i}}{\sqrt{2\pi} n_i^{n_i + 1/2}} \left(1 + \frac{n_i - 1}{g_i} \right)^{g_i} \left(1 - \frac{1}{g_i} \right)^{1/2 - g_i} \left\{ \left(1 + \frac{n_i - 1}{g_i} \right)^{n_i - 1/2} \right\}$$

$$- \prod_i \frac{g_i^{n_i}}{\sqrt{2\pi} n_i^{n_i + 1/2}} e^{n_i}. \qquad (6.72)$$

Here

$$\left(1 + \frac{n_i - 1}{g_i} \right)^{g_i} \left(1 - \frac{1}{g_i} \right)^{1/2 - g_i} = \left(1 - \frac{1}{g_i} \right)^{1/2} \left(\frac{1 + (n_i - 1)/g_i}{1 - 1/g_i} \right)^{g_i}$$

$$\simeq \left(1 - \frac{1}{g_i} \right)^{1/2} \left(1 + \frac{n_i - 1}{g_i} + \frac{1}{g_i} \right)^{g_i} = \left(1 - \frac{1}{g_i} \right)^{1/2} \left(1 + \frac{n_i}{g_i} \right)^{g_i}. \qquad (6.73)$$

Then

$$\Delta \simeq \prod_i \frac{g_i^{n_i}}{\sqrt{2\pi} n_i^{n_i + 1/2}} \underbrace{\left(1 + \frac{n_i}{g_i} \right)^{g_i}}_{\simeq e^{n_i}} \left\{ 1 + \frac{(n_i - 1)(n_i - 1/2)}{g_i} \right\} \left(1 - \frac{1}{g_i} \right)^{1/2}$$

$$- \prod_i \frac{g_i^{n_i}}{\sqrt{2\pi} n_i^{n_i + 1/2}} e^{n_i}$$

$$\simeq \left(\prod_i \frac{g_i^{n_i}}{\sqrt{2\pi} n_i^{n_i + 1/2}} e^{n_i} \right) \left(\sum_i \frac{n_i(n_i - 3/2)}{g_i} \right) = \tilde{W} \sum_i \frac{n_i(n_i - 3/2)}{g_i}. \qquad (6.74)$$

***See *e.g.* R. Courant [14], Vol. I, p.175:

$$e^N = \lim_{n \to \infty} \left(1 + \frac{N}{n} \right)^n.$$

Example 6.7: Free energy, entropy and latent heat on quantum statistics
Taking the number of states per particle of a perfect gas, in momentum range dp and volume V, to be $4\pi V p^2 dp/h^3$, show that at "high" temperatures the free energy F and the entropy S are given by

$$F \simeq -NkT\left[1 + \ln\left\{\frac{V}{N}\left(\frac{2\pi mkT}{h^2}\right)^{3/2}\right\}\right], \tag{6.75}$$

and

$$S \simeq Nk\left[\frac{3}{2}\ln T + \ln\frac{V}{N} + \frac{5}{2} + \frac{3}{2}\ln\left(\frac{2\pi mk}{h^2}\right)\right]. \tag{6.76}$$

Show that this result gives a reasonable expression for the latent heat of evaporation per atom.

Solution: We use Eq. (6.57) together with the Stirling approximation, *i.e.*

$$F = F_{MB} + kT\ln N! \simeq F_{MB} + NkT(\ln N - 1), \quad F_{MB} \stackrel{(5.20)}{=} -NkT\ln\sum_i g_i e^{-\epsilon_i/kT}. \tag{6.77}$$

Therefore:

$$
\begin{aligned}
F &= -NkT\ln\sum_i g_i e^{-\epsilon_i/kT} + NkT(\ln N - 1) = -NkT\left[1 + \ln\sum_i g_i e^{-\epsilon_i/kT} - \ln N\right] \\
&= -NkT\left[1 + \ln\sum_i \frac{4\pi V p_i^2 dp_i e^{-\epsilon_i/kT}}{h^3 N}\right] \\
&= -NkT\left[1 + \ln\frac{4\pi V}{h^3 N} + \ln\int_0^\infty p^2 e^{-p^2/2mkT}dp\right] \\
&\stackrel{(2.20)}{=} -NkT\left[1 + \ln\frac{4\pi V}{h^3 N} + \ln\frac{1}{4}\sqrt{2^3 m^3 \pi k^3 T^3}\right] = -NkT\left[1 + \ln\frac{V}{h^3 N} + \ln(2mkT\pi)^{3/2}\right] \\
&= -NkT\left[1 + \ln\left\{\frac{V}{N}\left(\frac{2\pi mkT}{h^2}\right)^{3/2}\right\}\right]. \tag{6.78}
\end{aligned}
$$

Correspondingly we have for the entropy:

$$S \stackrel{(6.55)}{=} S_{MB} - k\ln N! \simeq S_{MB} - kN(\ln N - 1), \tag{6.79}$$

and hence with $E = (3/2)NkT$ (see Eq. (6.60)) the entropy in the high temperature domain is

$$
\begin{aligned}
S \stackrel{(6.57)}{=} \frac{E}{T} - \frac{F}{T} &= \frac{E}{T} + Nk\left[1 + \ln\left\{\frac{1}{N}\sum_i g_i e^{-\epsilon_i/kT}\right\}\right] \\
&= Nk\left[\frac{3}{2} + 1 + \ln\left(\frac{V}{N}\right) + \frac{3}{2}\ln T + \frac{3}{2}\ln\left(\frac{2\pi mk}{h^2}\right)\right] \\
&= Nk\left[\frac{5}{2} + \ln\left(\frac{V}{N}\right) + \frac{3}{2}\ln T + \frac{3}{2}\ln\left(\frac{2\pi mk}{h^2}\right)\right]. \tag{6.80}
\end{aligned}
$$

With the relation (6.62) for L_{vapour}, the latent heat of evaporation per atom, we now obtain (as compared with the Maxwell–Boltzmann method of Example 5.4, Eq. (5.53)):

$$L_{\text{vapour}} \simeq \frac{S_{\text{vapour}}}{NT} = \frac{k}{T}\left[\frac{5}{2} + \ln\left(\frac{V}{N}\right) + \frac{3}{2}\ln T + \frac{3}{2}\ln\left(\frac{2\pi mk}{h^2}\right)\right]. \tag{6.81}$$

This expression is reasonable as it now implies a dependence of V on N, as one would expect.

Example 6.8: The rotational partition function of a diatomic molecule

Evaluate the rotational partition function Z_{rot} of a diatomic molecule of moment of inertia I semi-classically.

Solution: The partition function Z of a system of N particles was defined in Eq. (5.13) as

$$Z = \left(\sum_i g_i e^{-\epsilon_i/kT} \right)^N \equiv z^N. \tag{6.82}$$

For a single molecule of moment of inertia I the rotational energy is given by quantum mechanics as[†††]

$$\epsilon_J = \frac{J(J+1)h^2}{8\pi^2 I}, \qquad \text{with degeneracy} \quad g_J = 2J + 1, \tag{6.83}$$

and the rotational partition function z_{rot} is therefore

$$z_{\text{rot}} = \sum_J (2J+1) e^{-J(J+1)h^2/8\pi^2 IkT}. \tag{6.84}$$

Setting

$$x = \frac{h^2}{8\pi^2 IkT} \equiv \frac{\hbar^2}{2IkT}, \qquad \epsilon_J = J(J+1)xkT, \tag{6.85}$$

this becomes

$$z_{\text{rot}} = \sum_{J=0}^{\infty} (2J+1) e^{-J(J+1)x} = 1 + 3e^{-2x} + 5e^{-6x} + 7e^{-12x} + \cdots. \tag{6.86}$$

As in the classical result (3.25) of Example 3.6 we have $d\epsilon = (2J+1)kTxdJ$, and therefore we can rewrite z_{rot} semi-classically as the integral

$$z_{\text{rot}} = \int_0^{\infty} (2J+1) e^{-J(J+1)x} dJ = -\frac{1}{x} \int_0^{\infty} dJ \frac{d}{dJ} \left[e^{-J(J+1)x} \right] = \frac{1}{x} = \frac{8\pi^2 IkT}{h^2}. \tag{6.87}$$

With the *Euler–Maclaurin summation formula* (4.30) additional higher order terms can be obtained to provide better approximations; thus

$$z_{\text{rot}} = \frac{1}{x} + \frac{1}{3} + \frac{x}{15} - \frac{4x^2}{315} + \cdots. \tag{6.88}$$

Since the diatomic molecule vibrates along its axis, there is also a vibrational partition function (multiplying z_{rot}), and hence there are corresponding contributions to the thermodynamical functions.[‡‡‡]

[†††]Recall the relative Schrödinger equation of the 2-particle system separated in spherical coordinates (r, θ, φ):

$$\left[\frac{p_r^2}{2m_0} + \frac{l^2}{2m_0 r^2} + V(r) \right] \psi(r, \theta, \varphi) = \epsilon \psi(r, \theta, \varphi),$$

where the reduced mass is $m_0 = m_1 m_2/(m_1 + m_2)$. Here $l^2|\psi> = l_0(l_0 + 1)\hbar^2|\psi>$, $l_0 \equiv J$. See H.J.W. Müller–Kirsten [47], pp.205, 210.

[‡‡‡]See D.C. Mattis and R.H. Swendsen [44], pp.90-94. Both types of contributions are summarized in Appendices 2 and 3 of R.P.H. Gasser and W.G. Richards [28].

Example 6.9: Similar and dissimilar diatomic molecules

The energy ϵ of a rotating diatomic molecule,[§§§] considered as a rigid rotator with a single moment of inertia I, is $p_\varphi^2/2I$, where p_φ is the angular momentum. By quantizing this angular momentum according to the rules of the *Old Quantum Theory* (*i.e.* that of Bohr), show that

$$\epsilon_J = \frac{J^2\hbar^2}{2I}, \quad \text{where} \quad J = 0, 1, 2, \dots . \tag{6.89}$$

According to the proper quantum theory, *i.e.* here the Schrödinger equation, the correct result is

$$\epsilon_J = \frac{J(J+1)\hbar^2}{2I}, \quad \text{where} \quad J = 0, 1, 2, \dots, \tag{6.90}$$

if the two atoms are dissimilar, and each such energy has a multiplicity $g_J = 2J + 1$. Show from the relation (6.57) for the free energy in both Bose–Einstein and Fermi–Dirac statistics at high temperatures, *i.e.*

$$F = -NkT\left[1 + \ln\left\{\frac{1}{N}\sum_j g_j e^{-\epsilon_j/kT}\right\}\right], \tag{6.91}$$

that the rotational contribution to the entropy of a diatomic gas of N molecules is

$$S_{\text{rot}} = Nk\ln\left(\frac{2IkT}{\hbar^2}\right), \tag{6.92}$$

if the two atoms forming each molecule are dissimilar. [Hint: Replace the sums over J by integrals]. If the two atoms in each molecule are similar, only alternate levels could be occupied. Explain this and show that in this case

$$S_{\text{rot}} = Nk\ln\left(\frac{IkT}{\hbar^2}\right). \tag{6.93}$$

Solution: In the Old Quantum Theory

$$\oint p_\varphi d\varphi = n_\varphi h. \tag{6.94}$$

In the present problem $p_\varphi = \text{const.}$, and $n_\varphi \equiv J$. Therefore

$$p_\varphi \oint d\varphi = 2\pi p_\varphi = Jh, \quad p_\varphi = J\hbar, \quad \hbar = \frac{h}{2\pi}. \tag{6.95}$$

It follows that in this old and incorrect consideration the energy is

$$\epsilon_J = \frac{p_\varphi^2}{2I} = \frac{J^2\hbar^2}{2I}. \tag{6.96}$$

Next we set $a = 2Ik/\hbar^2$, and evaluate the partition function of one molecule with the correct expression of Eq. (6.90):

$$\sum_{J=0}^{\infty} g_J e^{-\epsilon_J/kT} = \sum_{J=0}^{\infty}(2J+1)e^{-J(J+1)/aT} \simeq \int_0^{\infty}(2J+1)e^{-J(J+1)/aT}dJ$$

$$= \left[-aTe^{-J(J+1)/aT}\right]_0^{\infty} = aT. \tag{6.97}$$

[§§§]Diverse aspects of diatomic gases are treated in D.J. Amit and Y. Verbin [2], pp.352-365, see also D. Chandler [12].

Hence the free energy F is

$$F = -NkT\left[1 + \ln\left(\frac{aT}{N}\right)\right] = -NkT - NkT\ln(aT) + NkT\ln(N). \qquad (6.98)$$

The entropy S is therefore given by

$$S = -\left(\frac{\partial F}{\partial T}\right)_V = Nk + Nk\ln(aT) - Nk\ln(N) + \frac{NkT}{aT}a$$

$$= 2Nk + Nk\ln(aT) - Nk\ln(N)$$

$$= 2Nk + Nk\ln\left(\frac{2IkT}{\hbar^2}\right) - Nk\ln(N). \qquad (6.99)$$

The rotational contribution to the entropy of the gas of N diatomic dissimilar (*i.e.* distinguishable) molecules is therefore

$$S_{\text{rot}} = Nk\ln\left(\frac{2IkT}{\hbar^2}\right). \qquad (6.100)$$

We now consider the case of a "mono-atomic diatomic" molecule, *i.e.* one made of identical elements. In general the total wave function Ψ_{tot} of a molecule consists of the product of wave functions describing different properties of the molecule, just as the total energy ϵ_{tot} of the molecule is the sum of the corresponding energy terms, *e.g.*

$$\epsilon_{\text{tot}} = \epsilon_{\text{trans}} + \epsilon_{\text{rot}} + \epsilon_{\text{vib}} + \epsilon_{\text{el}}, \qquad (6.101)$$

which is the sum of the translational energy part, the rotational energy part, the vibrational energy part, and the electronic energy part. The corresponding total wave function is the product

$$\Psi_{\text{tot}} = \psi_{\text{trans}} \times \psi_{\text{rot}} \times \psi_{\text{vib}} \times \psi_{\text{el}}, \qquad (6.102)$$

and the overall partition function is a corresponding product. Here we may be thinking of the simpler problem of the hydrogen atom. But in the case of the hydrogen molecule or the deuteron we would also have to take into account nuclear or *isospin* (the difference between neutron and proton), and hence there would be a wave function ψ_{ns} taking this nuclear spin into account (which, by multiplication, increases the number of states). As a case that immediately comes to mind consider a diatomic molecule. If the two atoms are dissimilar, the molecule is described as *hetero-atomic*, if they are identical the molecule is described as *mono-atomic* or *homonuclear*. The two cases differ in their symmetries, and this difference is described by a number called the *symmetry number* γ. This symmetry number is the number of indistinguishable positions into which the molecule can be rotated, and must be introduced into the quantitative description in order to avoid double counting.

If one considers the simplest case of zero nuclear spin (isospin), only one nuclear spin function exists for each of the two atoms and consequently also only one for the diatomic molecule, and this single function is symmetric. It follows that a homonuclear molecule in which the nuclei have zero nuclear spin must have a total wave function which is symmetric with respect to interchange of the nuclei or elements. Thus with the symmetric nuclear wave function must be associated a symmetric rotational wave function (implying angular momentum J even). However, since under these conditions (*i.e.* zero nuclear spin) no antisymmetric nuclear spin function exists, this cannot be multiplied by an antisymmetric rotational wave function. Thus the *complete absence of an antisymmetric wave function* implies the absence of antisymmetrical rotational states (J odd). Thus no molecules occupy odd rotational levels, and this implies the absence of alternate levels or spectral lines.[¶¶¶] This absence of alternate levels in the sum of Eq. (6.97) implies in the present case (*cf.* Example 6.12)

$$\sum_{J=0,2,4,\ldots} g_J e^{-\epsilon_J/kT} = \frac{aT}{2} = \frac{IkT}{\hbar^2}, \qquad (6.103)$$

[¶¶¶]For very detailed discussions see in particular J.E. Mayer and M.G. Mayer [45], pp.175-176, R.P.H. Gasser and W.G. Richards [28], p.117.

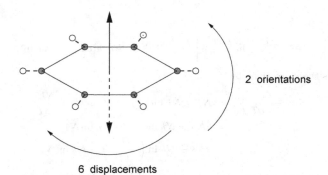

2 orientations

6 displacements

Fig. 6.10 Symmetry of the benzene molecule making the symmetry
factor $\gamma = 2 \times 6 = 12$.

and hence (compare with Eq. (6.100))

$$S_{\text{rot}} = Nk \ln \left(\frac{IkT}{\hbar^2} \right). \tag{6.104}$$

Thus only the fraction $1/\gamma$ of adjacent rotational levels have the correct symmmetry, where γ is the symmetry number referred to above. This number is the number of ways in which a molecule may be rotated into positions which would be different from the original one if the identical atoms were numbered and distinguishable, but which look identical to the original orientation in view of the identity of the atoms of the same element. For the diatomic molecule $\gamma = 2$.[****] Thus correspondingly in the case of the linear molecule CO_2 this number is $\gamma = 2$, in the case of tetrahedral methane CH_4 it is 12, as also in the case of planar hexagonal benzene C_6H_6, as illustrated in Fig. 6.10.[††††]

Example 6.10: Vibrational and electronic contributions to the entropy
Show that the *additional* contributions to the entropy S arising from any vibrational and electronic energy levels in a diatomic or polyatomic gas are

$$S_{\text{vib, elect}} = Nk \ln \sum_i g_i e^{-\epsilon_i/kT}, \tag{6.105}$$

where g_i and ϵ_i are respectively the relevant multiplicities (statistical weights) and energy eigenvalues — these would normally have to be determined from spectroscopic knowledge of the vibrational and electronic excitation levels.[‡‡‡‡]

Solution: The crux of this problem is that energies add but partition functions multiply (their logarithms add), as observed after Eq. (5.13). Therefore in the present case with i-th level energy ϵ_i of an element (molecule),

$$\epsilon_i = \epsilon_i^{\text{rot}} + \epsilon_i^{\text{vib}} + \epsilon_i^{\text{elec}}, \tag{6.106}$$

and degeneracies $g_i^{\text{rot}}, g_i^{\text{vib}}, g_i^{\text{elec}}$, the partition function Z is

$$Z = \prod (\sum_i g_i e^{-\epsilon_i/kT})^N = Z^{\text{rot}} Z^{\text{vib}} Z^{\text{elec}}, \tag{6.107}$$

[****] J.E. Mayer and M.G. Mayer [45], p.136.

[††††] J.E. Mayer and M.G. Mayer [45], p.195.

[‡‡‡‡] To avoid confusion: The contribution proportional to $N(\ln N - 1) \simeq \ln N!$ contained in the high T approximation (6.57) of the free energy F of both Bose–Einstein and Fermi–Dirac statistics is what distinguishes these correct expressions from that of Maxwell–Boltzmann statistics, Eqs. (5.11), (5.12). See comments after Eq. (6.52).

and the free energy F of Bose–Einstein and Fermi–Dirac statistics at high temperature T is

$$F \overset{(6.57)}{=} -NkT\left[1 - \ln N + \ln\sum_i g_i^{\text{rot}} e^{-\epsilon_i^{\text{rot}}/kT} + \ln\sum_i g_i^{\text{vib}} e^{-\epsilon_i^{\text{vib}}/kT} + \ln\sum_i g_i^{\text{elec}} e^{-\epsilon_i^{\text{elec}}/kT}\right],$$

(6.108)

and the entropy S is (*cf.* Eq. (6.53)), showing the additional contributions,

$$S = \frac{E}{T} + k\ln Z = \frac{E}{T} + Nk\left[1 - \ln N + \ln\sum_i g_i^{\text{rot}} e^{-\epsilon_i^{\text{rot}}/kT}\right.$$

$$\left. + \ln\sum_i g_i^{\text{vib}} e^{-\epsilon_i^{\text{vib}}/kT} + \ln\sum_i g_i^{\text{elec}} e^{-\epsilon_i^{\text{elec}}/kT}\right].$$

(6.109)

Example 6.11: The factors responsible for distinguishability

Show that the difference between the entropies calculated according to Maxwell–Boltzmann statistics in Example 5.4 on the one hand, and Bose–Einstein and Fermi–Dirac statistics at reasonably high temperatures on the other in Example 6.7, lies in the extra factors $N!/(n_1!)(n_2!)(n_3!)\ldots$ introduced into the number of arrangements in the Maxwell–Boltzmann case, *i.e.* in the factor expressing the distinguishability of the elements.

Solution: This aspect has effectively already been pointed out in the Examples cited. At fairly high temperatures the probability of occupation of any one energy level is very small, so that $n_i! \simeq 1$ for all i. Also, $\ln N! \sim N\ln N - N$, since N is a given large number. Finally, we recall that $S = k\ln(\text{number of arrangements})$.

6.9 Problems without Worked Solutions

Example 6.12: Absence of alternate levels

Prove the result (6.103), *i.e.* show that

$$\sum_{J=0,2,4,\ldots}^{\infty} e^{-J(J+1)/aT} = \frac{1}{2}aT.$$

(6.110)

Example 6.13: Number of states in a spherical volume V

Consider free nonrelativistic motion of particles in a spherical volume of radius R. What is the number of states with momentum p in the interval $(p, p + dp)$?

Example 6.14: Classical diatomic perfect gas

Consider a perfect gas consisting of diatomic molecules. Write down the Lagrangian function L of a single molecule of moment of inertia I. Let the harmonic vibrations of the two atoms along the axis and about an equilibrium position η_0 be described by the coordinate $x(t) = \eta(t) - \eta_0$, and the sum of kinetic and potential parts of the vibrations by $L_{\text{vib}} = (m/2)(\dot{x}^2 - \omega^2 x^2)$, where m_0 is the reduced mass of the two atoms and ω a constant. Write down the Hamilton function and count the number of quadratic terms. Then obtain the specific heat C_V per particle in this purely classical consideration with the help of the law of equipartition of energy. (Answer: $C_V = 7k/2$).

Example 6.15: Pressure of a particle in a cube

Consider a particle of mass m in a cube of volume $V = L^3$ and calculate its pressure P on a wall. [Hint: From the Schrödinger equation (6.10) the energy E of the particle is given by the eigenvalue relation (6.13) as

$$E = \frac{\hbar^2 \pi^2}{2mL^2}(l^2 + m^2 + n^2).$$

(6.111)

In a change dL of a length L the particle of energy E performs work $dW = -dE$. Use (F meaning force) $dW = FdL$ and $P = F/L^2$]. (Answer: $P = 2E/3V$).

Example 6.16: Number of photons

Consider thermal radiation at a temperature of $1000°\,K$. What is the wavelength λ of the radiation? What is the number N of photons in a volume V ignoring polarization?
(Answers: $\lambda = 14 \times 10^{-6}\,m$, $N \simeq 10^{15}$).

Chapter 7

Exact Form of Distribution Functions

7.1 Introductory Remarks

In the preceding chapter we derived W, the *number of arrangements of elements* (particles, atoms, ...) in quantum statistics. We now proceed to take the subsidiary conditions into account with the help of the method of *Lagrangian multipliers* of Chapter 2. This enables us to obtain the occupation numbers n_i (*i.e.* the number of elements occupying the single particle level i of energy ϵ_i). The resulting expressions for n_i in the two types of *quantum statistics* — here derived by the *method of maximization* of the number of arrangements W — are known as *distribution functions*. In Chapter 12 we shall see that the same expressions can be obtained as *mean occupation numbers* $\bar{n}_i$.

7.2 Fermi–Dirac Occupation Numbers

We begin with the expression (6.44) giving the number of arrangements of particles in Fermi–Dirac statistics, *i.e.*

$$W_{FD} = \prod_i \frac{g_i!}{n_i!(g_i - n_i)!}. \tag{7.1}$$

Taking the logarithm we obtain

$$\ln W_{FD} = \sum_i \ln g_i! - \sum_i [\ln n_i! + \ln(g_i - n_i)!]. \tag{7.2}$$

97

With the Stirling approximation

$$\ln n! \simeq n(\ln n - 1)$$

we obtain

$$\ln W_{FD} \simeq \sum_i \ln g_i! - \sum_i [n_i(\ln n_i - 1) + (g_i - n_i)\{\ln(g_i - n_i) - 1\}]. \quad (7.3)$$

As in the Maxwell–Boltzmann case (*cf.* Eq. (4.7)) we differentiate this expression and equate the result to zero in order to determine the *maximum distribution*. Thus

$$0 = d\ln W_{FD} \simeq -\sum_i [dn_i(\ln n_i - 1 + 1) - dn_i\{\ln(g_i - n_i) - 1 + 1\}]$$

$$= -\sum_i \ln\left(\frac{n_i}{g_i - n_i}\right) dn_i. \quad (7.4)$$

We now have to take into account *two subsidiary conditions*. The first condition we have is

$$E = \sum_i n_i \epsilon_i, \quad \text{and so} \quad 0 = dE = \sum_i \epsilon_i dn_i. \quad (7.5)$$

We consider separately the cases of *nonconserved elements* and *conserved elements.*

1. *Nonconserved elements* (actually none of these are known for Fermi–Dirac states). By the method of *Lagrangian multipliers* we obtain from Eqs. (7.4) and (7.5) by comparison with Eqs. (2.31) to (2.34):

$$\ln\left(\frac{n_i}{g_i - n_i}\right) = -\mu\epsilon_i, \qquad \frac{n_i}{g_i - n_i} = e^{-\mu\epsilon_i}, \quad (7.6)$$

and hence

$$n_i = \frac{g_i e^{-\mu\epsilon_i}}{1 + e^{-\mu\epsilon_i}},$$

and hence

$$n_i = \frac{g_i}{e^{\mu\epsilon_i} + 1}. \quad (7.7)$$

2. *Conserved elements* (*i.e.* for instance electrons or positrons). In this case we have the further subsidiary condition;

$$N \text{ (known)} = \sum_i n_i, \quad (7.8)$$

and hence, differentiating, we get

$$0 = dN = \sum_i dn_i.$$

Again using the method of *Lagrangian multipliers*, we obtain

$$\ln \frac{n_i}{g_i - n_i} = -\mu\epsilon_i + \text{const.}, \quad \text{const.} \equiv \mu\epsilon_0 \qquad (7.9)$$

(it is normally more convenient to write the constant $= \mu\epsilon_0$). Rearranging factors this becomes

$$\frac{n_i}{g_i - n_i} = e^{-\mu\epsilon_i + \text{const.}} \equiv e^{-\mu(\epsilon_i - \epsilon_0)}, \qquad (7.10)$$

or

$$n_i = \frac{g_i e^{-\mu\epsilon_i + \text{const.}}}{1 + e^{-\mu\epsilon_i + \text{const.}}} = \frac{g_i e^{-\mu(\epsilon_i - \epsilon_0)}}{1 + e^{-\mu(\epsilon_i - \epsilon_0)}},$$

and hence

$$n_i = \frac{g_i}{A e^{\mu\epsilon_i} + 1} = \frac{g_i}{e^{\mu(\epsilon_i - \epsilon_0)/kT} + 1}, \qquad (7.11)$$

where $A \equiv e^{-\epsilon_0/kT}$ is a constant. This constant A is obtained from $N = \sum_i n_i$.

We observe that $0 \le f_i = n_i/g_i \le 1$, *i.e.* the probability of a degenerate state i being occupied by a fermion lies between 0 and 1, in accordance with Pauli's principle.

7.3 Bose–Einstein Occupation Numbers

In the case of Bose–Einstein statistics we begin with the expression (6.46) giving the number of arrangements of particles on this statistics, *i.e.*

$$W_{BE} = \prod_i \frac{(n_i + g_i - 1)!}{(g_i - 1)! n_i!}. \qquad (7.12)$$

Here we have

$$\ln W_{BE} = \sum_i [\ln(n_i + g_i - 1)! - \ln(g_i - 1)! - \ln(n_i)!].$$

Again we use

$$\ln n! \simeq n(\ln n - 1).$$

Then

$$\ln W_{BE} = \sum_i [(n_i + g_i - 1)\{\ln(n_i + g_i - 1) - 1\}]$$

$$- \sum_i [(g_i - 1)\{\ln(g_i - 1) - 1\} + n_i(\ln n_i - 1)]$$

$$= \sum_i [(n_i + g_i - 1)\{\ln(n_i + g_i - 1)\}]$$

$$- \sum_i [g_i \ln(g_i - 1) + n_i \ln n_i]. \tag{7.13}$$

Again differentiating, we obtain (only with dn_i's)

$$d\ln W_{BE} = \sum_i [dn_i\{\ln(n_i + g_i - 1) + 1\}] - \sum_i [0 + dn_i(1 + \ln n_i)]. \tag{7.14}$$

Thus

$$d\ln W_{BE} = \sum_i \ln\left(\frac{n_i + g_i - 1}{n_i}\right) dn_i \simeq \sum_i \ln\left(\frac{n_i + g_i}{n_i}\right) dn_i, \tag{7.15}$$

since $n_i + g_i \gg 1$. Therefore for maximization of W_{BE} we have as the first condition:

$$0 = d\ln W_{BE} = -\sum_i \ln \frac{n_i}{n_i + g_i} dn_i. \tag{7.16}$$

Again we have to consider *subsidiary conditions*. The first such condition is

$$E \text{ (known)} = \sum_i n_i \epsilon_i, \quad 0 = dE = \sum_i \epsilon_i dn_i. \tag{7.17}$$

We consider again the cases of *nonconserved* and *conserved* elements separately.

1. For *nonconserved elements* as before we use Lagrangian multipliers and obtain from Eqs. (7.16) and (7.17) the relation

$$\ln \frac{n_i}{n_i + g_i} = -\mu\epsilon_i, \tag{7.18}$$

so that

$$n_i = (n_i + g_i)e^{-\mu\epsilon_i}, \quad \text{or} \quad n_i = \frac{g_i e^{-\mu\epsilon_i}}{1 - e^{-\mu\epsilon_i}},$$

and hence for nonconserved elements like photons and phonons:

$$n_i = \frac{g_i}{e^{\mu\epsilon_i} - 1}. \tag{7.19}$$

2. For *conserved elements, e.g.* the helium atom and atoms containing an even number of elementary particles, we get similarly as in the Fermi–Dirac case with an extra constant A the result:

$$n_i = (n_i + g_i)\frac{e^{-\mu\epsilon_i}}{A}, \qquad n_i = \frac{g_i}{Ae^{\mu\epsilon_i} - 1}, \qquad A \equiv e^{-\mu\epsilon_0}. \qquad (7.20)$$

Actually, of course, we should have $g_i - 1$, wherever we get g_i; but g_i is large and therefore 1 may be neglected (a mathematical reason). But also, the maximum probability would not be sharp unless we have a large number of elements (a physical reason). It is normally better to write the results altogether in the form

$$n_i = \frac{g_i}{e^{\mu(\epsilon_i - \epsilon_0)} \begin{smallmatrix} +1 \\ -1 \\ 0 \end{smallmatrix}}, \qquad \text{where} \quad \begin{cases} + \text{ implies FD statistics,} \\ - \text{ implies BE statistics,} \\ 0 \text{ implies MB statistics.} \end{cases} \qquad (7.21)$$

We note that in the case of the Bose–Einstein distribution the probability of a degenerate state i being occupied lies between 0 and infinity, since $0 \leq f_i = n_i/g_i \leq \infty$.

Finally we come to the *identification of μ*. We note:

1. For any two systems in thermal equilibrium (*i.e.* in thermal contact) the parameter μ is the same for both. The proof is exactly the same as for the classical system considered in Sec. 4.3.3. This is true even if the systems in thermal contact contain elements obeying different statistics, provided the system is closed.

2. Take for one system a perfect gas thermometer with high dilution of particles (very few particles per unit volume). This obeys the gas law

$$P = NkT, \qquad (7.22)$$

where N is the number of particles per unit volume. Then we obtain as for classical statistics in Sec. 4.3.4

$$\mu = \frac{1}{kT}. \qquad (7.23)$$

Hence the final form of the *distribution function* is

$$n_i = \frac{g_i}{e^{(\epsilon_i - \epsilon_0)/kT} \pm 1}, \qquad (7.24)$$

where $\epsilon_0 = 0$ for nonconserved elements like photons and phonons.

7.4 Thermodynamical Functions

For the same reasons as in Sec. 5.2 we again take as *definition of entropy* S the Boltzmann formula

$$S = k \ln W. \tag{7.25}$$

Whence $F = E - TS$ gives the *free energy* F, to be derived below in Example 7.1, in the two quantum statistics respectively as

$$F_{BE,FD} = N\epsilon_0 \pm kT \sum_i g_i \ln[1 \mp e^{(\epsilon_0 - \epsilon_i)/kT}]. \tag{7.26}$$

In the case of classical Maxwell–Boltzmann statistics we defined a partition function Z by the relation (5.13). Here in the case of quantum statistics we define correspondingly the partition functions Z_{BE}, Z_{FD} for Bose–Einstein and Fermi–Dirac statistics by the equations

$$F_{BE,FD} = -kT \ln Z_{BE,FD}, \tag{7.27}$$

and

$$\ln Z_{BE,FD} = -N \ln w_0 \mp \sum_i g_i \ln(1 \mp w_0 z_i), \tag{7.28}$$

where $\epsilon_0 = kT \ln w_0$, $z_i = e^{-\epsilon_i/kT}$. We shall see later in Sec. 12.4.3 that the same expression is obtained by the Darwin–Fowler method as the leading terms of the partition function in quantum statistics, and the different roles played by w_0 here and ω_0 there become evident. In the present case, of course, $\epsilon_0 = 0$ for nonconserved elements. We note for later reference that

$$
\begin{aligned}
\frac{\partial F}{\partial \epsilon_0} &= N \pm \frac{kT}{kT} \sum_i g_i \frac{\mp e^{(\epsilon_0 - \epsilon_i)/kT}}{1 \mp e^{(\epsilon_0 - \epsilon_i)/kT}} \\
&= \sum_i n_i - \sum_i \frac{g_i}{e^{(\epsilon_i - \epsilon_0)/kT} \mp 1} \\
&\overset{(7.24)}{=} 0
\end{aligned}
\tag{7.29}
$$

is equivalent to the equation $N = \sum_i n_i$ (*e.g.* if the free energy is a minimum with respect to all parameters). We also observe that $\partial F/\partial N = \epsilon_0$.

7.5 Applications and Examples

Example 7.1: The free energy F in quantum statistics
Derive from $S = k \ln W, W = W_{FD}, W_{BE}$, and the thermodynamical relation

$$F = E - TS, \qquad S = \frac{E}{T} - \frac{F}{T}, \tag{7.30}$$

the free energy F in Fermi–Dirac and Bose–Einstein statistics.

Solution: Consider *Fermi–Dirac statistics*. In this case

$$W_{FD} = \prod_i \frac{g_i!}{n_i!(g_i - n_i)!},$$ (7.31)

and hence

$$\frac{S_{FD}}{k} = \ln W_{FD} = \sum_i \ln g_i! - \sum_i [\ln n_i! + \ln(g_i - n_i)!].$$ (7.32)

Using again the Stirling approximation for $n \gg 1$,

$$\ln n! \simeq n(\ln n - 1),$$

we obtain (the non-logarithmic terms dropping out)

$$\begin{aligned}
\frac{S_{FD}}{k} &\simeq \sum_i g_i(\ln g_i - 1) - \sum_i n_i(\ln n_i - 1) - \sum_i g_i\{\ln(g_i - n_i) - 1\} \\
&\quad + \sum_i n_i\{\ln(g_i - n_i) - 1\} \\
&\simeq -\sum_i n_i \ln\left(\frac{n_i}{g_i - n_i}\right) + \sum_i g_i \ln\left(\frac{g_i}{g_i - n_i}\right).
\end{aligned}$$ (7.33)

With the help of Eq. (7.10) this becomes (using $N = \sum_i n_i, E = \sum_i n_i \epsilon_i$)

$$\frac{S_{FD}}{k} = \frac{1}{kT}\sum_i n_i(\epsilon_i - \epsilon_0) + \sum_i g_i \ln\left(\frac{g_i}{g_i - n_i}\right) = \frac{E}{kT} - \frac{N\epsilon_0}{kT} + \sum_i g_i \ln\left(\frac{1}{1 - n_i/g_i}\right).$$ (7.34)

Now using Eq. (7.11) we obtain with $\mu = 1/kT$ of Eq. (7.23),

$$\begin{aligned}
\frac{S_{FD}}{k} &= \frac{E}{kT} - \frac{N\epsilon_0}{kT} + \sum_i g_i \ln \frac{1}{1 - 1/[e^{\mu(\epsilon_i - \epsilon_0)} + 1]} \\
&= \frac{E}{kT} - \frac{N\epsilon_0}{kT} + \sum_i g_i \ln\left\{e^{-(\epsilon_i - \epsilon_0)/kT}\left[e^{(\epsilon_i - \epsilon_0)/kT} + 1\right]\right\} \\
&= \frac{E}{kT} - \frac{N\epsilon_0}{kT} + \sum_i g_i \ln\left[1 + e^{-(\epsilon_i - \epsilon_0)/kT}\right],
\end{aligned}$$ (7.35)

so that the free energy in the Fermi–Dirac case is

$$F_{FD} = E - S_{FD}T = N\epsilon_0 - kT\sum_i g_i \ln\left[1 + e^{(\epsilon_0 - \epsilon_i)/kT}\right].$$ (7.36)

The derivation of F in *Bose–Einstein statistics*, F_{BE} — which differs in some signs — proceeds correspondingly. We have

$$W_{BE} = \prod_i \frac{(n_i + g_i - 1)!}{n_i!(g_i - 1)!},$$ (7.37)

so that (non-logarithmic terms dropping out)

$$
\begin{aligned}
\frac{S_{BE}}{k} &= \ln W_{BE} = \sum_i \ln(n_i + g_i - 1)! - \sum_i \ln n_i! - \sum_i \ln(g_i - 1)! \\
&\simeq \sum_i (n_i + g_i - 1)\{\ln(n_i + g_i - 1) - 1\} - \sum_i [n_i\{\ln n_i - 1\} + (g_i - 1)\{\ln(g_i - 1) - 1\}] \\
&= \sum_i n_i \ln\left(\frac{n_i + g_i - 1}{n_i}\right) + \sum_i (g_i - 1) \ln\left(\frac{n_i + g_i - 1}{g_i - 1}\right) \\
&\overset{(7.20)}{\simeq} \sum_i n_i \ln\left(e^{(\epsilon_i - \epsilon_0)/kT}\right) + \sum_i g_i \ln\left(\frac{n_i + g_i}{g_i}\right) \\
&= \frac{1}{kT}\sum_i n_i(\epsilon_i - \epsilon_0) + \sum_i g_i \ln\left(1 + \frac{n_i}{g_i}\right) \\
&\overset{(7.24)}{=} \frac{E - N\epsilon_0}{kT} + \sum_i g_i \ln\left[1 + \frac{1}{e^{(\epsilon_i - \epsilon_0)/kT} - 1}\right] = \frac{E - N\epsilon_0}{kT} + \sum_i g_i \ln\left[\frac{e^{(\epsilon_i - \epsilon_0)/kT}}{e^{(\epsilon_i - \epsilon_0)/kT} - 1}\right] \\
&= \frac{E - N\epsilon_0}{kT} + \sum_i g_i \ln\left[\frac{1}{1 - e^{(\epsilon_0 - \epsilon_i)/kT}}\right] \\
&= \frac{E - N\epsilon_0}{kT} - \sum_i g_i \ln\left[1 - e^{(\epsilon_0 - \epsilon_i)/kT}\right], \quad\quad\quad (7.38)
\end{aligned}
$$

and therefore the free energy in the Bose–Einstein case is

$$
\begin{aligned}
F_{BE} &= E - S_{BE}T = E - (E - N\epsilon_0) + kT\sum_i g_i \ln\left[1 - e^{(\epsilon_0 - \epsilon_i)/kT}\right] \\
&= N\epsilon_0 + kT\sum_i g_i \ln\left[1 - e^{(\epsilon_0 - \epsilon_i)/kT}\right]. \quad\quad\quad (7.39)
\end{aligned}
$$

Example 7.2: Deriving energy E from the Gibbs–Helmholtz equation

Show from the thermodynamical Gibbs–Helmholtz equation that the internal energy E is given in terms of the free energy F by the relation

$$
E = -T^2\left[\frac{\partial(F/T)}{\partial T}\right]_V. \quad\quad\quad (7.40)
$$

Verify this relation for Maxwell–Boltzmann, Bose–Einstein and Fermi–Dirac statistics by calculating E from F by means of it, and comparing with the values obtained from the relation

$$
E = \sum_i n_i \epsilon_i. \quad\quad\quad (7.41)
$$

Solution: We have the relations

$$
F = E - TS, \quad \text{or} \quad \frac{F}{T} = \frac{E}{T} - S, \quad \text{and} \quad C_V := \left(\frac{dQ}{dT}\right)_V = \left(\frac{\partial E}{\partial T}\right)_V, \quad\quad\quad (7.42)
$$

and $dQ = TdS = dE + PdV$. With the help of these relations we obtain:

$$
\left[\frac{\partial(F/T)}{\partial T}\right]_V = \left[\frac{\partial}{\partial T}\left(\frac{E}{T}\right)\right]_V - \left[\frac{\partial S}{\partial T}\right]_V, \quad\quad\quad (7.43)
$$

and

$$\left[\frac{\partial S}{\partial T}\right]_V = \left[\frac{1}{T}\frac{\partial E}{\partial T}\right]_V, \tag{7.44}$$

and hence

$$\left[\frac{\partial(F/T)}{\partial T}\right]_V = -\left(\frac{E}{T^2}\right)_V + \left[\frac{1}{T}\frac{\partial E}{\partial T}\right]_V - \left[\frac{\partial S}{\partial T}\right]_V, \tag{7.45}$$

so that

$$E = -T^2\left[\frac{\partial(F/T)}{\partial T}\right]_V. \tag{7.46}$$

1. *Maxwell–Boltzmann statistics*: In this case we have:

$$\frac{F_{MB}}{T} \stackrel{(5.12)}{=} -Nk\ln\sum_i g_i e^{-\epsilon_i/kT}. \tag{7.47}$$

Hence:

$$\left[\frac{\partial}{\partial T}\left(\frac{F_{MB}}{T}\right)\right]_V = -Nk\frac{\sum_i g_i(\epsilon_i/kT^2)e^{-\epsilon_i/kT}}{\sum_i g_i e^{-\epsilon_i/kT}} = -\frac{N}{T^2}\frac{\sum_i g_i\epsilon_i e^{-\epsilon_i/kT}}{\sum_i g_i e^{-\epsilon_i/kT}}, \tag{7.48}$$

and therefore, with Eq. (7.44),

$$E = -T^2\left[\frac{\partial}{\partial T}\left(\frac{F_{MB}}{T}\right)\right]_V = \frac{\sum_i Ng_i\epsilon_i e^{-\epsilon_i/kT}}{\sum_i g_i e^{-\epsilon_i/kT}}. \tag{7.49}$$

On the other hand:

$$E = \sum_i n_i\epsilon_i \stackrel{(5.5)}{=} \sum_i \epsilon_i\left(\frac{Ng_i e^{-\epsilon_i/kT}}{\sum_i g_i e^{-\epsilon_i/kT}}\right) = \frac{\sum_i Ng_i\epsilon_i e^{-\epsilon_i/kT}}{\sum_i g_i e^{-\epsilon_i/kT}}. \tag{7.50}$$

2. *Fermi–Dirac and Bose–Einstein statistics*: We have, with upper signs for Fermi–Dirac statistics, lower signs for Bose–Einstein statistics, the following relation:

$$n_i = \frac{g_i}{w_0^{-1}e^{\epsilon_i/kT} \pm 1}, \qquad w_0 = e^{\epsilon_0/kT}, \tag{7.51}$$

and therefore

$$E = \sum_i n_i\epsilon_i = \sum_i \left(\frac{g_i\epsilon_i}{w_0^{-1}e^{\epsilon_i/kT} \pm 1}\right). \tag{7.52}$$

We also have from Eqs. (7.36) and (7.39) the relations:

$$F_{FD,BE} - N\epsilon_0 = kT\sum_i\left\{\mp g_i\ln(1 \pm w_0 z_i)\right\},$$

and hence

$$\frac{F_{FD,BE} - N\epsilon_0}{T} = k\sum_i\left\{\mp g_i\ln\left(1 \pm w_0 e^{-\epsilon_i/kT}\right)\right\}. \tag{7.53}$$

It follows that (ϵ_0 on the right originates from $\partial w_0/\partial T$)

$$\frac{\partial}{\partial T}\left(\frac{F_{FD,BE} - N\epsilon_0}{T}\right) = k\sum_i\frac{-g_i w_0((\epsilon_i - \epsilon_0)/kT^2)e^{-\epsilon_i/kT}}{1 \pm w_0 e^{-\epsilon_i/kT}}, \tag{7.54}$$

and therefore

$$-T^2 \frac{\partial}{\partial T}\left(\frac{F_{FD,BE} - N\epsilon_0}{T}\right) = \sum_i \frac{g_i w_0(\epsilon_i - \epsilon_0)e^{-\epsilon_i/kT}}{1 \pm w_0 e^{-\epsilon_i/kT}} = \sum_i \frac{g_i(\epsilon_i - \epsilon_0)e^{-\epsilon_i/kT}}{w_0^{-1} \pm e^{-\epsilon_i/kT}}$$

$$= \sum_i \frac{g_i(\epsilon_i - \epsilon_0)}{w_0^{-1}e^{\epsilon_i/kT} \pm 1} = \sum_i n_i\epsilon_i - \sum_i n_i\epsilon_0 = E - N\epsilon_0,$$

and thus

$$-T^2 \frac{\partial}{\partial T}\left(\frac{F_{FD,BE}}{T}\right) = E. \tag{7.55}$$

Example 7.3: The pressure in all 3 statistics

Show that in all three statistics the pressure $P = -(\partial F/\partial V)_T$ exerted by a perfect gas is given by $PV = (2/3)E$, and that $N\epsilon_0$ is equal to the thermodynamical potential $E - TS + PV$.

Solution: We consider the free energy of Fermi–Dirac and Bose–Einstein statistics,

$$F_{FD,BE} = N\epsilon_0 \mp kT \sum_i g_i \ln\left[1 \pm e^{(\epsilon_0 - \epsilon_i)/kT}\right]. \tag{7.56}$$

From Eq. (7.29) we know that $\partial F/\partial \epsilon_0 = 0$. Thus in differentiating with respect to the volume V we can ignore ϵ_0 (with its V dependence). It remains to differentiate g_i. We use the continuum expression (3.9c) of g_i, i.e. $g = 4\pi V p^2 dp/h^3$, and therefore obtain from $F_{FD,BE}$ with $\epsilon = p^2/2m$, and $e^{\epsilon_0/kT} = \eta$ for the pressure P:

$$P = -\left(\frac{\partial F}{\partial V}\right)_T = -\frac{\partial F}{\partial \epsilon_0}\frac{\partial \epsilon_0}{\partial V} - \frac{\partial F}{\partial g}\frac{\partial g}{\partial V} = -(\mp)kT \int_0^\infty \frac{4\pi p^2 dp}{h^3} \ln[1 \pm \eta e^{-p^2/2mkT}]. \tag{7.57}$$

Setting

$$x = \frac{p}{\sqrt{2mkT}}, \tag{7.58}$$

we obtain

$$\frac{P}{kT} = \pm \frac{4\pi(2mkT)^{3/2}}{h^3} \int_0^\infty x^2 \ln[1 \pm \eta e^{-x^2}]dx. \tag{7.59}$$

Now, by partial integration,

$$\int_0^\infty x^2 \ln[1 \pm \eta e^{-x^2}]dx = \left[\ln[1 \pm \eta e^{-x^2}]\overbrace{\int^x x^2 dx}^{x^3/3} - \int^x \frac{dx(\pm)\eta e^{-x^2}(-2x)(x^3/3)}{(1 \pm \eta e^{-x^2})}\right]_0^\infty$$

$$= 0 \pm \frac{2\eta}{3}\int_0^\infty \frac{x^4 dx e^{-x^2}}{1 \pm \eta e^{-x^2}}. \tag{7.60}$$

The energy E is

$$E = \sum_i n_i\epsilon_i \overset{(7.24)}{=} \sum_i \frac{g_i\epsilon_i}{e^{(\epsilon_i - \epsilon_0)/kT} \pm 1} = \int_0^\infty \frac{4\pi V p^2 dp}{h^3} \frac{p^2/2m}{(1/\eta)e^{p^2/2mkT} \pm 1}$$

$$= \frac{4\pi V}{h^3}(2mkT)^{3/2}kT \int_0^\infty \frac{x^4 dx}{(1/\eta)e^{x^2} \pm 1}$$

$$= \frac{4\pi V}{h^3}(2mkT)^{3/2}kT \int_0^\infty \frac{\eta e^{-x^2} x^4 dx}{1 \pm \eta e^{-x^2}}. \tag{7.61}$$

Therefore we obtain with the help of Eq. (7.60)

$$E = \frac{4\pi V}{h^3}(2mkT)^{3/2}kT\frac{3}{2}\left[\pm\int_0^\infty x^2 \ln[1\pm\eta e^{-x^2}]dx\right], \qquad (7.62)$$

and therefore with Eq. (7.59)

$$PV = \frac{2}{3}E. \qquad (7.63)$$

For the Maxwell–Boltzmann case we obtained this result earlier in Eq. (5.60). Here we see that this same result for a perfect gas is also obtained with quantum statistics. Thus the mean pressure per particle, $\bar{p}$, is 2/3 times the particle's kinetic energy $mv^2/2$.[*]

Example 7.4: $N\epsilon_0$ is equal to the free enthalpy

Show that $N\epsilon_0$ is equal to the thermodynamical potential $F(T,V)+PV = E-TS+PV \equiv G(T,P)$ called *free enthalpy* or *Gibbs function* (note, *enthalpy* or *total heat* is defined as $H(S,P) = VP + E(S,V), dH(S,P) = VdP + PdV + dE = VdP + TdS$, also $G(T,P) = H(S,P) - TS$).

Solution: The result follows immediately from the formula for the free energy, Eq. (7.56), by observing that (*cf.* Eq. (7.57)) the second term on the right is $-PV$. Thus Eq. (7.56) can be rewritten in the form

$$F_{FD,BE} + PV = N\epsilon_0 = G(T,P). \qquad (7.64)$$

Example 7.5: Zero free enthalpy for systems of non-conserved elements

Prove that the most general condition for which $N\epsilon_0$ is equal to the thermodynamic potential G is that $F - N\epsilon_0 \propto V$. Show that under such conditions $PV = -(F - N\epsilon_0)$, and that the thermodynamic potential G of systems composed of non-conserved elements (such as photons or phonons) is then zero.

Solution: The result follows from Eq. (7.64).

Example 7.6: Energy differences between various statistics

Putting $\eta = e^{\epsilon_0/kT}$ and noting that the occupation numbers on quantum statistics are (BE, FD)

$$n_i = \frac{g_i}{(1/\eta)e^{\epsilon_i/kT}\mp 1} = \frac{g_i\eta e^{-\epsilon_i/kT}}{1\mp\eta e^{-\epsilon_i/kT}} = g_i\eta e^{-\epsilon_i/kT}\left\{1\pm\eta e^{-\epsilon_i/kT} + \eta^2 e^{-2\epsilon_i/kT}\pm\cdots\right\}, \qquad (7.65)$$

calculate the energy $E = \sum_i n_i\epsilon_i$ of a gas at temperatures just low enough for the differences between the three different statistics to become noticeable. Determine the corresponding gas-laws by using the relation $PV = (2/3)E$. Confirm the results by direct calculation of P from the free energy F, making a similar expansion of F in rising powers of η.

Solution: We return to Eq. (7.61) for the energy E and expand the denominator of the integral as in Eq. (7.65). This means

$$\begin{aligned}
E &= \frac{4\pi V}{h^3}(2mkT)^{3/2}kT\int_0^\infty \frac{\eta e^{-x^2}x^4 dx}{1\pm\eta e^{-x^2}} \\
&= \frac{4\pi V}{h^3}(2mkT)^{3/2}kT\eta\int_0^\infty x^4 e^{-x^2}dx\{1\mp\eta e^{-x^2} + \eta^2 e^{-2x^2}\pm\cdots\} \\
&\overset{(2.20)}{=} \frac{4\pi V}{h^3}(2mkT)^{3/2}kT\eta\left[\frac{3}{8}\sqrt{\pi}\mp\eta\frac{3}{8}\sqrt{\frac{\pi}{2^5}} + \eta^2\frac{3}{8}\sqrt{\frac{\pi}{3^5}} - \cdots\right] \\
&= \frac{4\pi V}{h^3}(2mkT)^{3/2}kT\eta\frac{3}{8}\sqrt{\pi}\left[1\mp\frac{\eta}{2^{5/2}} + \frac{\eta^2}{3^{5/2}}\mp\cdots\right].
\end{aligned} \qquad (7.66)$$

---

[*]See *e.g.* F. Reif [59], Secs. 7.13 and 9.13, G.H. Wannier [83], pp.182-184.

Using the result (7.63) we obtain

$$PV = \frac{2}{3}E = \frac{4\pi V}{h^3}(2mkT)^{3/2}kT\eta\frac{1}{4}\sqrt{\pi}\left[1 \mp \frac{\eta}{2^{5/2}} + \frac{\eta^2}{3^{5/2}} \mp \cdots\right]. \tag{7.67}$$

On the other hand, using the free energy expression (7.56),

$$F_{FD,BE} = N\epsilon_0 \mp kT\sum_i g_i \ln[1 \pm e^{(\epsilon_0 - \epsilon_i)/kT}], \tag{7.68}$$

we obtain the pressure P as in Eq. (7.59):

$$P = -\left(\frac{\partial F}{\partial V}\right)_T \stackrel{(7.59)}{=} \pm\frac{4\pi}{h^3}(2mkT)^{3/2}kT\int_0^\infty x^2 \ln[1 \pm \eta e^{-x^2}]dx, \tag{7.69}$$

and hence

$$\begin{aligned}
PV &= \pm\frac{4\pi V}{h^3}(2mkT)^{3/2}kT\int_0^\infty x^2\left[\pm\eta e^{-x^2} - \frac{1}{2}\eta^2 e^{-2x^2} \pm \frac{1}{3}(\eta e^{-x^2})^3 - \cdots\right]dx \\
&\stackrel{(2.20)}{=} \pm\frac{4\pi V}{h^3}(2mkT)^{3/2}kT\left[\pm\eta\frac{1}{4}\sqrt{\pi} - \frac{\eta^2}{2}\frac{1}{4}\sqrt{\frac{\pi}{2^3}} \pm \frac{\eta^3}{3}\frac{1}{4}\sqrt{\frac{\pi}{3^3}} - \cdots\right] \\
&= \frac{4\pi V}{h^3}(2mkT)^{3/2}kT\eta\frac{1}{4}\sqrt{\pi}\left[1 \mp \frac{\eta}{2^{5/2}} + \frac{\eta^2}{3^{5/2}} \mp \cdots\right]. \tag{7.70}
\end{aligned}$$

We see that the results (7.67), (7.70) agree.

7.6   Problems without Worked Solutions

Example 7.7: Fermi–Dirac distribution from detailed balancing

Assuming Pauli's principle, energy conservation and the principle of microscopic reversibility, also known as the principle of detailed balancing, (*i.e.* that the number of reactions $1 + 2 \rightarrow 3 + 4$ is the same as the number of reactions $3 + 4 \rightarrow 1 + 2$), show that the probability for a fermion to occupy state i is

$$f_i = \frac{1}{1 + e^{\alpha + \beta\epsilon_i}}, \tag{7.71}$$

where α and β are constants. [Hint: Consider first the probability of reactions $1 + 2 \leftrightarrows 3 + 4$, with $1 - f_i$ denoting the probability that no particle occupies state i]. What do you conclude for reactions $1 + 2 + 3 \leftrightarrows 4 + 5 + 6$?

Example 7.8: Bose–Einstein distribution from detailed balancing

Obtain the Bose–Einstein distribution function by repeating the calculations of Example 7.7 but without the Pauli principle.

Example 7.9: Energy of an ideal gas in a spherical volume V

What is the relation corresponding to Eq. (7.63), *i.e.* $PV = 2E/3$, in the case of an ideal gas enclosed in a spherical volume of radius R? [Hint: The radial part of the Schrödinger equation is solved in terms of spherical Bessel functions. The zeros of these functions are known and can be looked up in Tables of Mathematical Functions].[†]

[†]Taken from T. Fliessbach [26], Exercise 8.1.

Chapter 8

Application to Radiation (Light Quanta)

8.1 Introductory Remarks

We now consider the application of the foregoing to radiation, and this means to light quanta. Why do light quanta obey Bose–Einstein statistics?

1. Light waves are indistinguishable, which means they have no labels. Therefore quantum statistics applies and not classical statistics.

2. Two light waves are superpositions, *i.e.* they obey no exclusion principle, as electrons for example do. Hence they should obey Bose–Einstein rather than Fermi–Dirac statistics in which all particles are *labelled*.

3. There is a selection rule for emission or absorption of radiation which says*

$$\triangle l = \pm 1, \qquad (8.1)$$

where $lh/2\pi$ is the total orbital angular momentum. Hence angular momentum is removed by the photon. Therefore the photon has (intrinsic) angular momentum $h/2\pi$, by conservation of angular momentum (since l changed by 1). Therefore the photon has spin 1. All Fermi–Dirac elements have half integer spin, $1/2, 3/2, 5/2, \ldots$ *etc.* All known Bose–Einstein elements have zero or integer spin, $0, 1, 2, \ldots$[†]

---

*W. Heitler [32] has shown that a photon resulting from a 2^l-pole electric or magnetic transition carries away an angular momentum $l\hbar$ with respect to the origin to which the multipole is referred. See also E. Fermi [24], p.96.

†This connection, here stated without proof, is known as *Pauli's theorem*. The proof of this connection is a crucial aspect of the subject and involves field theory, which is beyond the scope of this text. See, however, a simplified attempt for the case of spin 0 in Example 8.11.

Therefore it follows that light quanta obey Bose–Einstein statistics.

4. The photon is a *nonconserved element*. Therefore

$$n_i = \frac{g_i}{e^{\epsilon_i/kT} - 1} \tag{8.2}$$

is the distribution law.

5. The energy ϵ of a photon of frequency ν is given by

$$\epsilon = h\nu. \tag{8.3}$$

Its momentum p is given by

$$p = \frac{h\nu}{c}. \tag{8.4}$$

The factor g is in the case of the photon

$$g = 2 \times \underbrace{\frac{4\pi V}{h^3} p^2 dp}_{a \; priori \; \text{probability per particle in the range p,p+dp}}$$

$$= 2 \times \frac{4\pi V \nu^2 d\nu}{c^3}. \tag{8.5}$$

Here we have to multiply by the factor 2 because there are two polarizations of the transverse wave.

8.2   Planck's Radiation Law

Consider a classical simple harmonic oscillator described by the Lagrangian

$$L(q, \dot{q}) = \frac{1}{2}m\dot{q}^2 - 2\pi^2 m\nu^2 q^2. \tag{8.6}$$

Its Euler–Lagrange equation is

$$m\ddot{q} + 4\pi^2 m\nu^2 q = 0$$

with solution

$$q(t) \propto \cos(2\pi\nu t + \text{phase}) \quad (\text{or sin}). \tag{8.7}$$

This solution represents a *mode of vibration*. Correspondingly we can have oscillators in orthogonal directions. Then every such harmonic oscillator represents a *normal mode of vibration*. Considering a system of such classical simple harmonic oscillators we can say: For each simple harmonic oscillator one normal mode of vibration is "emitted". Therefore a simple harmonic oscillator is *statistically equivalent* to a normal mode of vibration. As we saw in Chapter 3 (*cf.* Eq. (3.12)), the classical simple harmonic oscillator has an *a priori* probability of having energy between ϵ and $\epsilon + d\epsilon$ proportional to

$$\frac{d\epsilon}{\nu}, \tag{8.8}$$

where ν is the frequency of vibration which is the same as the frequency of the radiation given off. Furthermore, if we assume as in classical theory, that the energy levels are continuous, we arrive at the unacceptable answer, that the energy density (*i.e.* the total energy per unit volume) is infinite (this is the Rayleigh–Jeans law, which is considered in Example 8.2). For this reason Planck (1901) introduced the concept of *discrete energy levels*.[‡] Then the probability of having energy between ϵ and $\epsilon + d\epsilon$ is only approximately $d\epsilon/\nu$, and we have to replace the continuous energy distribution by a discrete energy distribution.

Therefore: Number of discrete energy levels per unit energy range $\propto$ *a priori* probability $\propto 1/\nu$ (*cf.* after Eq. (3.12)). Therefore generalizing to discrete energy levels (since number of levels $\propto$ *a priori* probability), we must have: Number of energy levels per unit energy range $\propto 1/\nu$, or — written differently —

$$\frac{dn}{d\epsilon} = \frac{\text{number of discrete energy levels}}{d\epsilon} \propto \frac{1}{\nu}. \tag{8.9}$$

Therefore if this is true, the spacing of these levels of an energy scale must be proportional to ν (*i.e.* this is — *cf.* Eq. (8.9) — $d\epsilon$ divided by the number of levels). Therefore the actual energy levels themselves are

$$\epsilon \propto n\nu, \quad n \text{ an integer.}$$

Therefore we write

$$\epsilon = nh\nu,$$

h being in 1901 an unknown constant.

Next we consider the average (mean) energy of a simple harmonic oscillator at a certain temperature T. We write this mean energy $\bar{\epsilon}$. We use classical statistics. Then the probability that the oscillator will have the energy $\epsilon = nh\nu$ is (with the Maxwell–Boltzmann distribution function)

$$\propto e^{-\epsilon/kT} = e^{-nh\nu/kT}.$$

[‡]See Example 8.3.

Hence the mean energy $\bar{\epsilon}$ is given by

$$\bar{\epsilon} = \frac{\sum_{n=0}^{\infty} nh\nu e^{-nh\nu/kT}}{\sum_{n=0}^{\infty} e^{-nh\nu/kT}}. \tag{8.10}$$

Here n is (so to speak) the variable. We write

$$\phi = \frac{h\nu}{kT}. \tag{8.11}$$

Then

$$\bar{\epsilon} = h\nu \frac{\sum_{n=0}^{\infty} ne^{-n\phi}}{\sum_{n=0}^{\infty} e^{-n\phi}}. \tag{8.12}$$

But the sum in the numerator is equal to $-\partial/\partial\phi$ of the sum in the denominator. Therefore

$$\bar{\epsilon} = -h\nu \frac{\partial}{\partial\phi} \ln\left[\sum_{n=0}^{\infty} e^{-n\phi}\right]. \tag{8.13}$$

This trick is extremely advantageous in statistical mechanics because it gets rid of arbitrary constants at an early stage. Hence (summing the geometric progression)

$$\bar{\epsilon} = -h\nu \frac{\partial}{\partial\phi} \ln \frac{1}{1 - e^{-\phi}} = h\nu \frac{\partial}{\partial\phi} \ln(1 - e^{-\phi})$$

$$= h\nu \frac{e^{-\phi}}{1 - e^{-\phi}} = \frac{h\nu}{e^{\phi} - 1} = \frac{h\nu}{e^{h\nu/kT} - 1}. \tag{8.14}$$

This is the mean energy of the simple harmonic oscillator and is equal to the mean energy of one normal mode of vibration of frequency ν in the radiation. Now, the number of normal modes of vibration between ν and $\nu + d\nu$ is

$$\frac{8\pi\nu^2 d\nu V}{c^3}.$$

Therefore the mean energy of radiation between frequencies ν and $\nu + d\nu$ is

$$= \frac{8\pi h V \nu^3 d\nu}{c^3(e^{h\nu/kT} - 1)} \overset{c=\nu\lambda}{=} \frac{8\pi h V c}{\lambda^5(e^{hc/\lambda kT} - 1)} d\lambda. \tag{8.15}$$

This result is also known as *Planck's distribution law*. Observe the denominator results here without reference to Bose–Einstein statistics. The examples in the last section explore and illuminate this result from various directions.

8.3   Black Body Thermal Radiation

In Example 8.6 it is shown that the total radiation density (*i.e.* that integrated over frequency or wavelength) depends only on the temperature T. Thus the radiation described by Planck's law is determined by T alone, independent of the body's shape or composition. Conversely one can argue that any macroscopic body or substance with temperature is a radiator, *i.e.* emits and/or absorbs radiation, which is therefore called *thermal radiation*. One cannot attribute a temperature to a microscopic particle since this cannot radiate thermally. Heat radiation is a macroscopic phenomenon.

An idealized physical body that absorbs all incident electromagnetic radiation, regardless of wavelength and incident direction, is called a *black body* (analogously a body that reflects similarly all incident radiation is described as a white body). A black body can also emit radiation. In fact, Planck's radiation law describes the radiation of a black body at constant temperature T in equilibrium with its surroundings (meaning its emission and absorption is such that the constant temperature T is maintained — which is not possible in the case of a microscopic particle).

An idealized model of a black body is the surface of a small hole in an insulated enclosure which acts as boundary between inside and outside. The radiation passing through the hole into the enclosure is continually reflected (or absorbed and emitted) by the surface inside with very little escaping to outside. In equilibrium at temperature T the energy density of the radiation is described by Planck's law. In general there may be absorption, reflection and transmission. For a perfectly black body there is no reflection or transmission. If the wavelength of the radiation is larger than the diameter of the hole the radiation cannot pass through.

The black hole considered in general relativity is a region of spacetime from which nothing escapes. The geometrical surface surrounding it is called *event horizon*. The inside domain is called black because all light hitting the event horizon is absorbed (escape from inside would require a velocity larger than the velocity of light c). Furthermore it has been shown in cosmology (see Chapter 14 here and L. Susskind and J. Lindesay [73], pp.48-49) that a large black hole appears to a distant observer as a body with temperature $T = 1/8\pi MG$ and energy M (in units with $c = 1, \hbar = 1$), G being Newton's gravitational constant. As a consequence one expects a thermal atmosphere to be associated with the horizon and energy to be carried away as thermal radiation. This process discovered by S. Hawking is referred to as *Hawking radiation*. The rate of energy loss L (called *luminosity*) can therefore be shown to be given by

$$L \propto T^4 \times area,$$

where the area is that of the horizon. This relation is essentially the Stefan–Boltzmann law of black body radiation as shown in Examples below and in Chapter 14.

8.4   Applications and Examples

Example 8.1: Number of modes of vibration

Show that the number of possible modes of vibration of radiation per unit volume with frequencies between ν and $\nu + d\nu$ is

$$\frac{8\pi\nu^2 d\nu}{c^3}, \tag{8.16}$$

where c is the velocity of light.

Solution: The number of possible modes of vibration per unit volume is given by

$$\frac{dp_x dp_y dp_z}{h^3}. \tag{8.17}$$

Here

$$p_x^2 + p_y^2 + p_z^2 = p^2, \quad p = \frac{h\nu}{c},$$

the latter being the momentum for radiation. Hence $dp = h d\nu/c$. In spherical coordinates a spherical volume element is given by the expression $dp_x dp_y dp_z = 4\pi p^2 dp$. Hence the number of possible modes of vibration per unit volume in the interval between frequencies ν and $\nu + d\nu$ is given by — remembering that light has two possible transverse polarizations —

$$\frac{dp_x dp_y dp_z}{h^3} \rightarrow \frac{4\pi p^2 dp}{h^3} \times 2 = \frac{4\pi h^2 \nu^2 h d\nu \times 2}{h^3 c^2 . c} = \frac{8\pi\nu^2 d\nu}{c^3}. \tag{8.18}$$

Example 8.2: Rayleigh–Jeans law by unacceptable continuum arguments

Assuming the statistical equivalence of waves and harmonic oscillators, show from the law of equipartition of energy valid for an *energy continuum* in Maxwell–Boltzmann statistics that the energy density per unit frequency range would be the *Rayleigh–Jeans law*

$$E(\nu) = \frac{8\pi\nu^2 kT}{c^3}. \tag{8.19}$$

Show that this scheme, based on an energy continuum, is not acceptable.

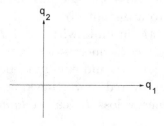

Fig. 8.1 Two possible directions of polarization.

Solution: The energy of the radiation wave is proportional to the sum of the oscillator contributions

$$2 \text{ kinetic terms } + 2\pi^2 m\nu^2 q_1^2 + 2\pi^2 m\nu^2 q_2^2, \tag{8.20}$$

where m is the equivalent mass, ν is the frequency, and q_1, q_2 are the displacements in the two possible directions of polarization as indicated in Fig. 8.1, *i.e.* when $q_1 = q_{1,\max}, q_2 = 0$ and conversely. Therefore the mean energy for each mode of vibration gives $\frac{1}{2}kT + \frac{1}{2}kT = kT$. Hence — since the number of possible modes (or states) per unit volume per unit frequency range is $= 8\pi\nu^2/c^3$ according to Example 8.1, we obtain

$$E(\nu) = \frac{8\pi\nu^2}{c^3} \times kT. \tag{8.21}$$

This scheme based on an energy continuum is not acceptable because it cannot represent the distribution of energy throughout the spectrum of a black body, for at a given temperature T, the total radiation density $\mathcal{E} = \int E(\nu)d\nu$ tends to infinity as ν tends to infinity, or the wavelength λ to zero. Only at long λ, *i.e.* small ν, is the above expression roughly correct.

Example 8.3: Planck's law from Maxwell–Boltzmann statistics

Show from Maxwell–Boltzmann statistics applied to the discrete energy levels of an oscillator that the mean oscillator energy is

$$\bar{\epsilon} = \frac{h\nu}{e^{h\nu/kT} - 1}, \tag{8.22}$$

and hence deduce the *Planck radiation law* (energy per unit volume and unit interval of frequency)

$$E(\nu) = \frac{8\pi h\nu^3}{c^3(e^{h\nu/kT} - 1)}. \tag{8.23}$$

Solution: The Maxwell–Boltzmann distribution law for nonconserved elements like photons is with $g_i = \Delta q \Delta p/h = \Delta\epsilon/h\nu$, as was shown in the text (*cf.* Eq. (4.10) and the lines after Eq. (3.12) with $d\epsilon \to \Delta\epsilon$), the expression

$$n_i = g_i e^{-\epsilon_i/kT} = \frac{\Delta\epsilon}{h\nu} e^{-\epsilon_i/kT}. \tag{8.24}$$

Here $\epsilon_i = nh\nu, \Delta\epsilon = h\nu$, so that

$$n_i = g_i e^{-\epsilon_i/kT} = e^{-nh\nu/kT}. \tag{8.25}$$

It follows that the number of oscillators N is

$$N = \sum_i n_i = \sum_{n=0}^{\infty} e^{-nh\nu/kT} = \frac{1}{1 - e^{-h\nu/kT}}. \tag{8.26}$$

For the energy E of the radiation one obtains[§]

$$E = \sum_i n_i \epsilon_i = \sum_{n=1}^{\infty} e^{-nh\nu/kT} nh\nu = h\nu \sum_{n=1}^{\infty} n e^{-nh\nu/kT}. \tag{8.27}$$

[§]This must have been the observation of Planck: The sums with $\epsilon_i \to \epsilon_n, \epsilon_n = nh\nu$ are easy to evaluate.

But

$$\frac{1}{(1-x)^2} = 1 + 2x + 3x^2 + \cdots = \sum_{n=1}^{\infty} n x^{n-1}.$$

Therefore

$$E = h\nu e^{-h\nu/kT} \sum_{n=1}^{\infty} n e^{-(n-1)h\nu/kT} = \frac{h\nu e^{-h\nu/kT}}{(1 - e^{-h\nu/kT})^2}. \tag{8.28}$$

The mean energy $\bar{\epsilon}$ of an oscillator is therefore given by

$$\bar{\epsilon} = \frac{E}{N} = \frac{h\nu e^{-h\nu/kT}}{(1 - e^{-h\nu/kT})^2} \frac{(1 - e^{-h\nu/kT})}{1} = \frac{h\nu e^{-h\nu/kT}}{1 - e^{-h\nu/kT}} = \frac{h\nu}{e^{h\nu/kT} - 1}. \tag{8.29}$$

Now classically from Eq. (8.21):

$$E(\nu) = \frac{8\pi\nu^2 kT}{c^3},$$

but from Example 8.1:

$$\text{number of states per unit volume between } \nu \text{ and } \nu + d\nu = \frac{8\pi\nu^2 d\nu}{c^3}.$$

Therefore the energy density $E(\nu)$ which is equal to the number of oscillators per unit volume ($V = 1$) in unit frequency range ($d\nu = 1$) times the mean oscillator energy ($\bar{\epsilon}$) is

$$E(\nu) = \frac{8\pi h\nu^3}{c^3(e^{h\nu/kT} - 1)}. \tag{8.30}$$

Example 8.4: Planck's radiation law from Bose–Einstein statistics

Derive the Planck radiation law by considering the radiation to be represented by an unprescribed number of photons of energy $\epsilon = h\nu$ obeying Bose–Einstein statistics.

Solution: In Bose–Einstein statistics of non-conserved elements:

$$n_i \overset{(7.19)}{=} \frac{g_i}{e^{\epsilon_i/kT} - 1}, \qquad \epsilon_i = h\nu, \qquad \therefore n = \frac{g}{e^{h\nu/kT} - 1}. \tag{8.31}$$

Hence in the present case the energy E of the radiation is

$$E = \sum_i n_i \epsilon_i = \sum g \frac{h\nu}{e^{h\nu/kT} - 1} = \int \frac{dp_x dp_y dp_z}{h^3} V \frac{h\nu}{e^{h\nu/kT} - 1}$$

$$\overset{(8.18)}{=} \int \frac{8\pi\nu^2 d\nu}{c^3} V \frac{h\nu}{e^{h\nu/kT} - 1}. \tag{8.32}$$

Hence

$$\frac{dE}{V} \equiv \frac{E(\nu)d\nu}{V} = \frac{8\pi h\nu^3}{c^3(e^{h\nu/kT} - 1)} d\nu, \tag{8.33}$$

as in Eq. (8.30).

Example 8.5: Wien displacement law

From the Planck law, derive the *Wien displacement law*

$$\lambda_m T = \text{constant}, \tag{8.34}$$

where λ_m is the wavelength for which the energy density is greatest. Show that the value of the constant is $hc/4.965k$.

Solution: Setting $x = hc/\lambda kT$, the Planck distribution law (8.15) is proportional to

$$f(\lambda)d\lambda = \frac{1}{\lambda^5(e^{hc/\lambda kT} - 1)}d\lambda. \tag{8.35}$$

The wavelength λ_m of the maximum of $f(\lambda)$ is obtained by equating to zero the derivative of $f(x)$, *i.e.* as the solution of

$$0 = -5\lambda + \frac{hc}{kT(e^{hc/\lambda kT} - 1)}e^{hc/\lambda kT}, \quad i.e. \quad -5(e^x - 1) + xe^x = 0, \quad i.e. \quad e^x\left(1 - \frac{x}{5}\right) = 1. \tag{8.36}$$

This equation can be solved graphically, but one can also check that $x = 4.965$ solves the equation. Thus the wavelength λ_m and frequency ν_m at the maximum are given by

$$\lambda_m T = \frac{hc}{4.965k}, \quad \nu_m/T = \frac{4.965k}{h}. \tag{8.37}$$

The behaviour of the Planck distribution function as a function of frequency is shown in Fig. 8.2.

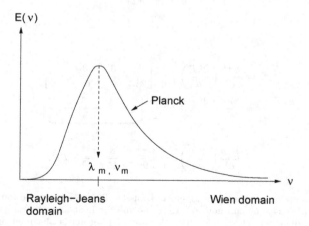

Fig. 8.2 The Planck distribution law as a function of frequency.

Example 8.6: Stefan–Boltzmann law

By integrating $E(\nu)$ (of Eqs. (8.23), (8.30)) over all frequencies ν, show that the *total* radiation density $\mathcal{E}$ is proportional to T^4 (Stefan–Boltzmann law), and evaluate the constant of proportionality.

Solution: We have from Example 8.3:[¶]

$$\mathcal{E} = \int_0^\infty E(\nu)d\nu = \int_0^\infty \frac{8\pi h\nu^3 d\nu}{c^3(e^{h\nu/kT} - 1)} = \int_0^\infty \frac{e^{-h\nu/2kT}8\pi h\nu^3 d\nu}{c^3(e^{h\nu/2kT} - e^{-h\nu/2kT})}. \tag{8.38}$$

This would give infinity with $E(\nu)$ of Eq. (8.21). Setting

$$x = \frac{h\nu}{2kT}, \quad d\nu = \frac{2kT}{h}dx, \tag{8.39}$$

we obtain

$$\begin{aligned}
\mathcal{E} &= \frac{8\pi h}{c^3}\left(\frac{2kT}{h}\right)^4 \int_0^\infty \frac{e^{-x}x^3 dx}{e^x - e^{-x}} = \frac{8\pi h}{c^3}\left(\frac{2kT}{h}\right)^4 \int_0^\infty \frac{e^{-2x}x^3 dx}{1 - e^{-2x}} \\
&= \frac{8\pi h}{c^3}\left(\frac{2kT}{h}\right)^4 \int_0^\infty e^{-2x}x^3 dx\left[1 + e^{-2x} + (e^{-2x})^2 + \cdots\right].
\end{aligned} \tag{8.40}$$

Next we set

$$x = u^2 \quad \text{and} \quad dx = 2udu, \quad x^3 dx = 2u^7 du.$$

Then

$$\mathcal{E} = \frac{8\pi h}{c^3}\left(\frac{2kT}{h}\right)^4 \int_0^\infty 2e^{-2u^2}u^7\left[1 + e^{-2u^2} + e^{-4u^2} + \cdots\right]du. \tag{8.41}$$

The integrals of this series can be evaluated with the help of the Gaussian integral (2.20), *i.e.*

$$\int_0^\infty u^7 e^{-\alpha u^2} du = \frac{3}{\alpha^4}. \tag{8.42}$$

Then

$$\begin{aligned}
\mathcal{E} &= \frac{8\pi h}{c^3}\left(\frac{2kT}{h}\right)^4 \times 2\left[\frac{3}{2^4} + \frac{3}{4^4} + \frac{3}{6^4} + \cdots\right] \\
&= \frac{8\pi h}{c^3}\left(\frac{2kT}{h}\right)^4 \times \frac{6}{2^4}\underbrace{\left[1 + \frac{1}{2^4} + \frac{1}{3^4} + \cdots\right]}_{\pi^4/90} \\
&= \frac{8\pi h}{c^3}\left(\frac{2kT}{h}\right)^4 \times \frac{3}{8}\frac{\pi^4}{90} = \frac{8}{15}\frac{\pi^5}{c^3 h^3}(kT)^4,
\end{aligned} \tag{8.43}$$

where in the last step we used a result given in Tables.[‖] This result can be found in all standard books.[**] With Eq. (5.24) one can now calculate the specific heat of a photon gas.[††] As compared with the pressure of a perfect gas, Eq. (7.63), that of photons is obtained as $P = \mathcal{E}/3V$; *cf.* Example 8.9.

[¶]For the evaluation of Bose–Einstein integrals see also R.B. Dingle [17], pp.98-99.
[‖]H.B. Dwight [23], formula 48.5, p.11.
[**]F. Reif [59], Sec. 9.13, D.C. Mattis and R.H. Swendsen [44], p.104.
[††]D.J. Amit and Y. Verbin [2], p.447.

Example 8.7: Radiation pressure

Show that in the case of photons in a volume V the equation of state is

$$PV = \frac{1}{3}E. \tag{8.44}$$

Solution: The derivation proceeds as in Example 7.3 but we short-circuit this here for simplicity by appealing to the result (7.63) for a perfect gas. In the case of a photon the energy is $\epsilon = \tilde{p}c$, where $\tilde{p}$ is its momentum and c is the velocity of light. In the case of a free molecule of mass m the energy ϵ is $\epsilon = \tilde{p}^2/2m = \tilde{p}v/2$, where v is the velocity. Thus, expressed in terms of the momentum $\tilde{p}$ and with $N/V = \bar{n}$ the mean number of molecules in volume V, the pressure given by Eq. (7.63) is $P = 2\bar{n}\epsilon/3 = \bar{n}\tilde{p}v/3$. Therefore in the case of photons we have

$$P = \frac{1}{3}\bar{n}\tilde{p}c = \frac{1}{3}\bar{n}\epsilon = \frac{E}{3V}. \tag{8.45}$$

8.5   Problems without Worked Solutions

Example 8.8: Stefan–Boltzmann law from thermostatics

Using the relation $TdS = dE + PdV$ with $E = Vu$, where u is the mean energy density of radiation, and using the result (8.45) for the pressure of a photon gas, show that $u(T) \propto T^4$.[‡‡] [Hint: Eliminate S with the help of Eq. (1.13)].

Example 8.9: Stefan–Boltzmann law from Carnot cycle

In Example 8.7 it was shown that the radiation pressure P in a volume V is as follows related to the mean energy density $\psi = E/V$:

$$P = \frac{1}{3}\frac{E}{V} = \frac{1}{3}\psi, \qquad dP = \frac{1}{3}d\psi. \tag{8.46}$$

Consider a quasistatically working Carnot machine consisting of a piston in a cylinder with a small hole at the bottom (the latter replacing the usual thermal contact of the Carnot cycle), both being frictionless, thermally insulated and perfectly reflecting. The hole can be covered by a similarly perfect plate. An isothermally kept hollow space or cavity with temperature $T_1 = T + dT$ (the source) is attached on the other side of the hole, so that radiation can pass through the hole into the cylinder. A second similar such hollow space (acting as sink) has temperature $T_2 = T$. Describe an infinitesimal Carnot cycle in which the machine performs quasistatically the mechanical work dW and absorbs the amount of radiation (heat) Q_1 and gives off the amount Q_2. The efficiency of the machine is defined as the ratio dW/Q_1. Explain why

$$\frac{dW}{Q_1} = \frac{T_1 - T_2}{T_1} = \frac{dT}{T}, \tag{8.47}$$

and deduce from this the Stefan–Boltzmann law $E \propto T^4$. [Hints: In step (1): Work performed in pushing the piston: $(dW)_1 = (1/3)\psi_1(V_2 - V_1)$, energy added to fill the volume: $(dE)_1 = \psi_1(V_2 - V_1)$, $Q_1 = (4/3)\psi_1(V_2 - V_1)$, where ψ_1 is the energy density at $T = T_1$. In step (2): Reduction of temperature and ψ: $dP = (1/3)d\psi$. In step (3): Machine now opposite second hollow at temperature T_2 and hole opened: $Q_2 = (dE)_2 + \int_{V_3}^{V_4} PdV$. In step (4): adiabatic compression to (P_1, V_1). Total work done $= (P, V)$-area covered by the cycle, and $Q_1 - Q_2 = dW$. Reversibility requires $dS_1 + dS_2 = 0$, $Q_1/T_1 = Q_2/T_2$].

[‡‡]This law is of considerable generality. Thus a black hole loses mass at a rate proportional to T^4 times area. See Example 14.5

Example 8.10: Equilibrium between pair creation and radiation

(a) Rewriting the distribution function (7.24) for Fermi–Dirac and Bose–Einstein statistics respectively in the form

$$\frac{g_i \mp n_i}{n_i} = e^{(\epsilon_i - \epsilon_0)/kT}, \tag{8.48}$$

show from energy conservation of a reaction $1 + 2 \rightleftharpoons 1' + 2'$ (the numbers representing particles) that

$$n_1 n_2 (g_{1'} \mp n_{1'})(g_{2'} \mp n_{2'}) = n_{1'} n_{2'} (g_1 \mp n_1)(g_2 \mp n_2). \tag{8.49}$$

(b) Consider photons 1 and 2 as creating an electron–positron pair, each with mass m and energy respectively ϵ_- and ϵ_+. The reverse process is also possible, and thus assume both types of processes maintain equilibrium by detailed balancing between electron–positron creation and blackbody radiation. Show that the chemical potentials of the electron and positron vanish.[§§]

Example 8.11: The spin-statistics connection (simplified case)

Show that an anticommutation quantization condition for a scalar (*i.e.* spin-zero) field (also called *Klein–Gordon field*) leads to a contradiction. A convenient way to start is to consider a model theory of a scalar field $\varphi(x^\mu)$. A quantity φ defined with respect to some space (*e.g.* Euclidean, Minkowskian) is a *scalar* if it transforms like a scalar in that space (see *e.g.* [48], p.134). Thus consider the following Lagrangian density $\mathcal{L}$[¶¶]

$$\mathcal{L} = -\frac{\partial \varphi}{\partial x_\mu} \frac{\partial \varphi^*}{\partial x^\mu} - m^2 \varphi \varphi^*. \tag{8.50}$$

(a) Show that the variation with respect to φ^* as in classical mechanics yields the *Klein–Gordon equation*,

$$(\Box - m^2)\varphi = 0, \tag{8.51}$$

where $\Box = \partial_\mu \partial^\mu$. Also show that with $\Pi(\mathbf{x}) = \partial\mathcal{L}/\partial\dot\varphi = \dot\varphi^*$, the Hamilton density $\mathcal{H}$ is

$$\mathcal{H} = \Pi\Pi^* + \partial_\mu \varphi^* \partial^\mu \varphi + m^2 \varphi^* \varphi \geq 0. \tag{8.52}$$

As equal time $(t = t')$ quantization condition of the Klein–Gordon field one can take (in analogy to ordinary quantum mechanics, with $[A, B] = AB - BA$):

$$[\Pi(\mathbf{x}), \varphi(\mathbf{x}')] = [\dot\varphi^*(\mathbf{x}), \varphi(\mathbf{x}')] = -i\delta(\mathbf{x} - \mathbf{x}'), \tag{8.53a}$$

or

$$[\Pi^*(\mathbf{x}), \varphi^*(\mathbf{x}')] = [\dot\varphi(\mathbf{x}), \varphi^*(\mathbf{x}')] = -i\delta(\mathbf{x} - \mathbf{x}'). \tag{8.53b}$$

At different times one first writes (observe that φ, φ^* are independent operators like q and p in ordinary quantum mechanics) the operator relation as

$$[\varphi(\mathbf{x}, t), \varphi^*(\mathbf{x}', t')] = -i\Delta(\mathbf{x} - \mathbf{x}', t - t'), \tag{8.54}$$

[§§]See also P.T. Landsberg [38], problems 10.4, 13.5 on pp.253, 413.
[¶¶]See *e.g.* S. Schweber [68], p.190.

and determines the quantity $\triangle$, also called *Schwinger function* [69], from the conditions (observe $(\square - m^2)\triangle = 0)^{***}$

$$\triangle(\mathbf{x} - \mathbf{x}', 0) = 0, \qquad \frac{\partial \triangle(\mathbf{x} - \mathbf{x}', t - t')}{\partial t}\bigg|_{t=t'} = -\delta(\mathbf{x} - \mathbf{x}'). \qquad (8.55)$$

Verify these conditions (since $\triangle$ diverges on the light cone it should be considered as a distribution).
(b) Could we take on the left hand side of Eq. (8.54) not the commutator with minus sign but the anticommutator with plus sign? Show that this plus sign leads to a contradiction by setting $\varphi = a + ib \neq 0, \varphi^* = a - ib$ (with $a = a^*, b = b^*$). Hence the scalar field φ obeys Bose–Einstein statistics (not Fermi–Dirac statistics).
[Hint: Use $\triangle = 0$ for $\mathbf{x} = \mathbf{x}', t = t'$, so that

$$\varphi(\mathbf{x}, t)\varphi^*(\mathbf{x}, t) + \varphi^*(\mathbf{x}, t)\varphi(\mathbf{x}, t) = 0$$

implying $a = 0, b = 0$, hence $\varphi = 0$, contradicting the assumption $\varphi \neq 0$].

***See S. Schweber [68], p.180.

Chapter 9

Debye Theory of Specific Heat of Solids

9.1 Introductory Remarks

In the following we consider solids in the form of crystals which have a regular atomic structure, as indicated in Fig. 9.1. The atoms in the crystal interact with each other. Clearly, in order to be able to investigate properties of such structures quantitatively, one has to distinguish between dominant and minor aspects. A primary approximation is to consider only nearest-neighbour interactions of the atoms, as indicated by springs in Fig. 9.1, and to assume that these interactions are effectively oscillations or vibrations of the atoms around their equilibrium positions, and hence sound waves. These vibrations affect, of course, the entire crystal lattice. Clearly a first and simpler model of the crystal is that which ignores the interactions and assumes independent oscillations of the individual atoms. This can be further simplified by assuming that all the atoms oscillate with the same frequency ν like one-dimensional simple linear harmonic oscillators. This is the simplest model to begin with, and is considered in detail in Example 9.2. The model is known as the *Einstein model*. Since a one-dimensional linear harmonic oscillator is easy to handle both classically and quantum mechanically, such a model can be treated quantum mechanically.

A *phonon* is a quantized sound wave.* Phonons obey Bose–Einstein statistics for the same reasons as before for photons which are quantized light waves.

---

*Phonons can therefore be identified with elastic waves or pressure waves of ordinary sound or *first sound*. But a sound wave propagated in a phonon gas is different from these phonons, and is therefore described as *second sound*. See *e.g.* S.K. Ma [42], p.103 or R.B. Dingle [21], p.116. Second sound waves may be considered as temperature oscillations. See *e.g.* L.E. Reichl [57], p.538.

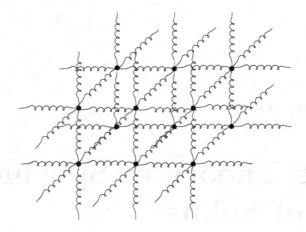

Fig. 9.1 Atomic crystal structure with nearest neighbour interactions.

Phonons are *nonconserved* elements since, as in the case of photons, their number in a volume V cannot be controlled in view of phonons leaving or entering at the boundary of the solid of volume V. Hence to relate their case to that of the preceding with the previous theory we replace the velocity of light by the velocity of sound. However, there is one marked difference between sound waves and light waves: Sound waves are longitudinal waves and hence there is only one polarization. But in solids we can also have two transverse polarizations, and hence there are really *three waves* operating here. (In solids the forces of cohesion are very great).

1. Hence we start from the approximation of a *continuum* (no atoms, *i.e.* perfectly uniform). In this approximation sound waves of all frequencies can be excited. But solids <u>do</u> consist of atoms, and these are spaced more or less equally, and it is quite impossible to pass through those wavelengths which are less than the distance between atoms, *cf.* Fig. 9.2, *i.e.* to propagate lattice vibrations with wavelengths less than the distance between atoms.

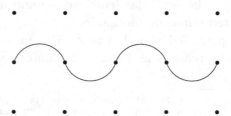

Fig. 9.2 The shortest wave which has any meaning.

Therefore the shortest wavelength is of the order of $2\times$ the interatomic distance.

2. It is quite wrong to consider all possible frequencies. We must cut off the high frequencies which correspond to small wavelengths $\lambda < 2\times$ the interatomic distance.

We now work this out in the next section.

9.2 The Calculation

Let u_l be the velocity of the longitudinal sound wave (of compression) for which there is *one* polarization, and let u_t be the velocity of the transverse (acoustical) waves for which there are *two* polarizations.[†] Then the number of *normal modes of vibration* with frequencies between ν and $\nu + d\nu$ is (in analogy to the photon case of Eq. (8.5))

$$d\mathcal{N} = 4\pi V \left(\frac{1}{u_l^3} + \frac{2}{u_t^3} \right) \nu^2 d\nu \equiv aV\nu^2 d\nu, \quad \text{say.} \qquad (9.1)$$

If we now integrate this over all ν we would get infinity. But (*cf.* the Dulong–Petit law of Sec. 4.4.2) for each direction, the energy of motion of each atom has two squared terms (momentum and displacement for each direction). For N atoms the total number of squared terms in the energy is $6N$, because there are 3 directions of motion with three momentum squared terms (*i.e.* 3 terms of linear translation) and 3 squared terms involving coordinates per atom (of harmonic oscillations in three directions about the equilibrium position of an atom). Therefore in total, there are $3N$ normal modes (*i.e.* periodic functions solving the $3N$ equations of motion) in the total system. Therefore the upper limit $\nu_{\max}$ is given by the relation

$$\int_0^{3N} d\mathcal{N} = aV \int_0^{\nu_{\max}} \nu^2 d\nu = 3N. \qquad (9.2)$$

Therefore:

$$aV\nu_{\max}^3 = 9N, \quad aV = \frac{9N}{\nu_{\max}^3}, \qquad (9.3)$$

or

$$\nu_{\max} = \left(\frac{9N}{aV} \right)^{1/3}. \qquad (9.4)$$

[†]See *e.g.* D.C. Mattis and R.H. Swendsen [44], p.105.

Therefore ν_{max} depends only on the velocities and the total number of particles in the system. This corresponds to the rough physical argument given above.

In order to be able to write down an expression for the total energy E of the crystal of N atoms we recall first (*cf. e.g.* Eq. (8.14)) that the mean energy of an oscillator (representing nonconserved elements, *i.e.* here phonons) of frequency ν in a heat bath of temperature T is

$$\bar{\epsilon}(\nu) = \frac{h\nu}{e^{h\nu/kT} - 1}. \tag{9.5}$$

Multiplying this expression by $d\mathcal{N}$, the number of oscillators or normal modes of vibration with frequency in the range $\nu, \nu + d\nu$, and integrating over all permissible values of the frequency ν, we obtain the total energy E given by[‡]

$$E = aV \int_0^{\nu_{max}} \frac{h\nu}{e^{h\nu/kT} - 1} \nu^2 d\nu. \tag{9.6}$$

This is the total energy since $h\nu/(e^{h\nu/kT} - 1)$ is the energy for each normal mode of vibration, *i.e.* the mean energy of each normal mode of vibration as obtained from Bose–Einstein statistics or from Planck's argument. We set

$$x = \frac{h\nu}{kT}, \qquad \nu = \frac{xkT}{h}. \tag{9.7}$$

Then

$$E = aV \int_0^{x_{max}} \frac{x^3 k^3 T^3}{h^3} h \frac{kT}{h} \frac{dx}{e^x - 1} = \frac{aV(kT)^4}{h^3} \int_0^{x_{max}} \frac{x^3 dx}{e^x - 1}. \tag{9.8}$$

The *Debye characteristic temperature* θ is defined as

$$\theta := \frac{h\nu_{max}}{k}, \qquad \therefore \quad x_{max} = \frac{h\nu_{max}}{kT} = \frac{\theta}{T}. \tag{9.9}$$

We now write the answer in terms of θ and substitute for aV from Eq. (9.3). Then

$$E = \frac{9N}{\nu_{max}^3} \frac{(kT)^4}{h^3} \int_0^{x_{max}} \frac{x^3 dx}{e^x - 1} = \frac{9Nh^3}{k^3\theta^3} \frac{(kT)^4}{h^3} \int_0^{x_{max}} \frac{x^3 dx}{e^x - 1}$$

$$= \frac{9NkT^4}{\theta^3} \int_0^{\theta/T} \frac{x^3 dx}{e^x - 1}. \tag{9.10}$$

[‡]This ignores an additive constant E_0 representing the ground state energy of the lattice.

The *high temperature approximation* is defined by the condition $T \gg \theta$. Therefore in this case θ/T is small, therefore x is always small and $x \ll 1$. Hence we subsitute for e^x the approximation

$$e^x - 1 \simeq x,$$

and obtain

$$\int \frac{x^3 dx}{x} = \frac{x^3}{3}.$$

It follows that the energy becomes:

$$E \to \frac{3NkT^4}{\theta^3} \frac{\theta^3}{T^3} = 3NkT. \tag{9.11}$$

We recognize this result as the Dulong–Petit answer, already contained in Eq. (4.29). The specific heat C is obtained as

$$C = \frac{\partial E}{\partial T} = 3Nk. \tag{9.12}$$

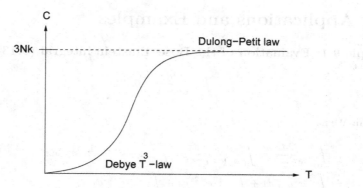

Fig. 9.3 The variation of E with θ.

At *low temperatures* $T \ll \theta$, and therefore θ/T is very large. Hence we approximate the upper limit by infinity. Then[§]

$$\int_0^\infty \frac{x^3 dx}{e^x - 1} = \frac{\pi^4}{15} = 6.494, \tag{9.13}$$

so that

$$E = \frac{9NkT^4}{\theta^3} \frac{\pi^4}{15} = \frac{3\pi^4 NkT^4}{5\theta^3}. \tag{9.14}$$

[§]For the evaluation see Example 9.1.

It follows that

$$C = \frac{\partial E}{\partial T} = \frac{12\pi^4 N k T^3}{5\theta^3}. \tag{9.15}$$

This result is the *Debye T^3 law*, valid at low temperatures only, as illustrated in Fig. 9.3. For comparison of Debye's theory with experiment see *e.g.* Mandl [41], pp.165-166.

In between the two limiting temperature domains a more complex law holds for the *Debye temperature* (*cf.* Eqs. (9.9) and (9.4)):

$$\theta = \frac{h}{k} \left[\frac{9N/V}{4\pi(1/u_l^3 + 2/u_t^3)} \right]^{1/3}, \tag{9.16}$$

or *Debye frequency* $\omega_D = k\theta/\hbar$. This gives the *lattice specific heat, i.e.* that due to vibrations of the atoms in the crystal or metal. This expression varies enormously because it depends on u_l and u_t. For soft substances $\theta \simeq 100°K$, and for hard substances $\theta \simeq 400°K$ (from this expression). The result (9.16) can be found, for instance, in the text of Amit and Verbin.[¶]

9.3   Applications and Examples

Example 9.1: Evaluation of the Bose–Einstein integral (9.13)
Show that

$$\int_0^\infty \frac{x^3 dx}{e^x - 1} = 6.494.$$

Solution: We have:

$$\int_0^\infty \frac{x^3 dx}{e^x - 1} = \int_0^\infty \frac{x^3 e^{-x} dx}{1 - e^{-x}} = \int_0^\infty (1 - e^{-x})^{-1} x^3 e^{-x} dx$$

$$= \int_0^\infty x^3 e^{-x} dx + \int_0^\infty x^3 e^{-2x} dx + \int_0^\infty x^3 e^{-3x} dx + \cdots$$

$$= \int_0^\infty x^3 e^{-x} dx + \frac{1}{2^4} \int_0^\infty (2x)^3 e^{-2x} d(2x) + \frac{1}{3^4} \int_0^\infty (3x)^3 e^{-3x} d(3x) + \cdots$$

$$= 6\left(1 + \frac{1}{2^4} + \frac{1}{3^4} + \cdots\right) = \frac{6 \times \pi^4}{90} = \frac{\pi^4}{15} = 6.494, \tag{9.17}$$

where we used a result given in Tables.[‖]

Example 9.2: A crystal treated as $3N$ linear harmonic oscillators
Consider a crystal of N atoms as one element with possible energy levels E_I. The possible modes of vibration are distinguishable from one another by virtue of their location and orientation within the crystal — for the atoms are localized and therefore distinguishable, and this distinguishability

---

[¶]D.J. Amit and Y. Verbin [2], p.383.
[‖]H.B. Dwight [23], formula 48.5, p.11.

will be preserved on transformation to a system of normal coordinates. Hence the appropriate statistics are Maxwell–Boltzmann statistics, for which the free energy F is (*cf.* Eq. (5.12))

$$F_{MB} = -kT \ln \left[\sum_{\mathcal{I}} g_{\mathcal{I}} e^{-E_{\mathcal{I}}/kT} \right]. \tag{9.18}$$

Treating the crystal as represented by $3N$ linear harmonic oscillators, each oscillator having possible energy values $E_i = (n + 1/2) h \nu_i$, where $n = 0, 1, 2, \ldots$, show that

$$I_{3N}(1) := \sum_{\mathcal{I}} g_{\mathcal{I}} e^{-E_{\mathcal{I}}/kT} = \left(e^{-h\nu_i/2kT} \sum_{n=0}^{\infty} e^{-nh\nu_i/kT} \right)^{3N} = \left(\frac{e^{-h\nu_i/2kT}}{1 - e^{-h\nu_i/kT}} \right)^{3N}. \tag{9.19}$$

Here $e^{-E_{\mathcal{I}}/kT}$ is the energy weighting factor for a specific combination of the positive integers n_i (which may be thought of as the number of phonons in level i, each with energy $h\nu$). Calculate the free energy, entropy, energy, zero-point energy and specific heat of the crystal.

Solution: As given, the crystal consists of a lattice of N atoms, each of these atoms having three independent components of motion about its equilibrium position and each individual motion being approximated by that of a one-dimensional linear harmonic oscillator. With these assumptions, the crystal of N atoms is thus represented by a system of $3N$ independent linear harmonic oscillators. To simplify this model called *Einstein model*** further, it is assumed that all $3N$ oscillators have the same frequency ν, an assumption which is improved in the theory of Debye. Each of the $3N$ individual oscillators is characterized by a quantum number $n_i = 0, 1, 2, \ldots$ for $i = 1, 2, \ldots, 3N$, so that the energy of the entire crystal is given by

$$E_{\mathcal{I}} = E(n_1, n_2, n_3, \ldots, n_{3N}) = E_1 + E_2 + E_3 + \cdots + E_{3N} \equiv E(n_1) + E(n_2) + \cdots + E(n_{3N})$$

$$= h\nu \left(n_1 + n_2 + n_3 + \cdots + n_{3N} + \frac{1}{2} 3N \right). \tag{9.20}$$

Thus

$$e^{-E_{\mathcal{I}}/kT} = e^{-h\nu(n_1 + n_2 + n_3 + \cdots + n_{3N} + 3N/2)/kT}$$

$$= e^{-h\nu(n_1 + 1/2)/kT} e^{-h\nu(n_2 + 1/2)/kT} e^{-h\nu(n_3 + 1/2)/kT} \cdots e^{-h\nu(n_{3N} + 1/2)/kT}.$$

Since the levels of the one-dimensional harmonic oscillator are nondegenerate, each such exponential on the right appears only once (*cf.* Eq. (12.16)). Since all combinations of specific values of $n_1, n_2, n_3, \ldots, n_{3N}$ are possible, one defines as the *partition function of the system* (actually in a heat bath of temperature T) the function Z given by the *sum over all possible arrangements* $\mathcal{I}$ of the integers n_i as

$$Z := \sum_{\mathcal{I}} g_{\mathcal{I}} e^{-E_{\mathcal{I}}/kT} \equiv \sum_{n_1, n_2, n_3, \ldots, n_{3N}} e^{-h\nu(n_1 + n_2 + n_3 \cdots + n_{3N} + 3N/2)/kT} \equiv (z)^{3N}, \tag{9.21}$$

where z is the *single particle partition function*

$$z = e^{-h\nu/2kT} \sum_{n=0}^{\infty} e^{-h\nu n/kT} = \frac{e^{-h\nu/2kT}}{1 - e^{-h\nu/kT}}, \tag{9.22}$$

where the sum is an ordinary geometric progression. The free energy F of the system is then (with factor g_i later in Eq. (12.18) equal to 1)

$$F = -kT \ln Z = -3NkT \ln z. \tag{9.23}$$

**D.J. Amit and Y. Verbin [2], p.248, T.L. Hill [34], pp.86-92.

The quantity

$$\theta_E := \frac{h\nu}{k} \tag{9.24}$$

is called *Einstein temperature*. In contrast with the above, the Debye approximation improves the simple Einstein-model assumption of identical frequencies of all oscillators by considering the mean energy of an oscillator of energy $h\nu$ in a heat bath of temperature T. The latter when multiplied by the number of such oscillators in the domain $\nu, \nu + d\nu$, and integrated over all permissible frequencies gives the total energy of the crystal of N atoms.

In the present case of the simpler model we obtain the free energy F now from Eq. (5.12), *i.e.* $F = -kT \ln Z$, as

$$F = -3NkT \ln z \overset{(9.22)}{=} 3N \left[\frac{1}{2} h\nu + kT \ln \left(1 - e^{-h\nu/kT} \right) \right]. \tag{9.25}$$

The entropy S is obtained with the help of Eqs. (5.10) and (5.12),

$$S = \frac{E}{T} + k \ln Z.$$

Here E is obtained from Eq. (5.23), *i.e.*

$$\begin{aligned}
E &= -\frac{\partial \ln Z}{\partial (1/kT)} = -\frac{3N \partial \ln z}{\partial (1/kT)} \\
&= -3N \frac{\partial}{\partial (1/kT)} \left[-\frac{h\nu}{2kT} - \ln \left(1 - e^{-h\nu/kT} \right) \right] \\
&= 3N \left(\frac{1}{2} h\nu + \frac{h\nu}{e^{h\nu/kT} - 1} \right).
\end{aligned} \tag{9.26}$$

The first term on the right is clearly the contribution of the zero point energy. Hence

$$\begin{aligned}
S &= \frac{3N}{T} \left(\frac{1}{2} h\nu + \frac{h\nu}{e^{h\nu/kT} - 1} \right) + \left(-\frac{F}{T} \right) \\
&= \frac{3N}{T} \left(\frac{1}{2} h\nu + \frac{h\nu}{e^{h\nu/kT} - 1} \right) - \frac{3N}{T} \left[\frac{1}{2} h\nu + kT \ln \left(1 - e^{-h\nu/kT} \right) \right] \\
&= \frac{3N}{T} \left[\frac{h\nu}{e^{h\nu/kT} - 1} - kT \ln \left(1 - e^{-h\nu/kT} \right) \right].
\end{aligned} \tag{9.27}$$

The specific heat C_V is obtained from Eq. (5.25):

$$\begin{aligned}
C_V &= T \left(\frac{\partial S}{\partial T} \right)_V \\
&= -\frac{3N}{T} \left[\frac{h\nu}{e^{h\nu/kT} - 1} - kT \ln \left(1 - e^{-h\nu/kT} \right) \right] \\
&\quad + 3N \left[\frac{-h\nu(-h\nu/kT^2)}{(e^{h\nu/kT} - 1)^2} e^{h\nu/kT} - k \ln \left(1 - e^{-h\nu/kT} \right) - \frac{kT(h\nu/kT^2)}{(1 - e^{-h\nu/kT})} \left(-e^{-h\nu/kT} \right) \right] \\
&= -\frac{3Nh\nu}{T} \frac{1}{e^{h\nu/kT} - 1} + 3Nk \left(\frac{h\nu}{kT} \right) \left[\left(\frac{h\nu}{kT} \right) \frac{e^{h\nu/kT}}{(e^{h\nu/kT} - 1)^2} + \frac{1}{(e^{h\nu/kT} - 1)} \right] \\
&= 3Nk \left(\frac{h\nu}{kT} \right)^2 \frac{e^{h\nu/kT}}{(e^{h\nu/kT} - 1)^2}.
\end{aligned} \tag{9.28}$$

This result and the Einstein model are discussed in more detail in the book of Amit and Verbin [2], pp.243-252, and p.306.

Example 9.3: A crystal treated as N 3-dimensional isotropic oscillators

Show that in Example 9.2 it would have been possible to consider the crystal as N 3-dimensional isotropic harmonic oscillators with possible energy levels

$$E_i = \left(n_x + n_y + n_z + \frac{3}{2}\right)h\nu_i \equiv \left(n + \frac{3}{2}\right)h\nu_i, \quad n = n_x + n_y + n_z = 0, 1, 2, \ldots, \quad (9.29)$$

instead of as $3N$ *linear* harmonic oscillators with possible energy levels $\epsilon_i = (n + 1/2)h\nu_i$. What is the value of $e^{-E_I/kT}$ now?

Solution: We have to evaluate the sum in Eq. (9.19), *i.e.*

$$\sum_I g_I e^{-E_I/kT}. \quad (9.30)$$

Here, with all frequencies identical for simplicity,

$$E_I = \left(n_1 + n_2 + n_3 + \cdots + n_N + \frac{3}{2}N\right)h\nu. \quad (9.31)$$

The degeneracy of the levels of a single 3-dimensional isotropic harmonic oscillator has been cited in Eq. (6.7) and is derived in Example 12.1; the result is the factor

$$g = \frac{1}{2}(n + 1)(n + 2) \quad \text{for} \quad E_n = \left(n + \frac{3}{2}\right)h\nu. \quad (9.32)$$

Thus we have

$$I_N(3) := \sum_I g_I e^{-E_I/kT} = \left(e^{-3h\nu/2kT} \sum_{n=0}^{\infty} \frac{1}{2}(n + 1)(n + 2)e^{-nh\nu/kT}\right)^N. \quad (9.33)$$

Is this expression equal to that in Eq. (9.19)? In order to see this we consider the cubic of the geometric progression

$$1 + x + x^2 + x^3 + \cdots \quad \text{with} \quad x = e^{-h\nu/kT},$$

i.e.

$$\left(\sum_{n=0}^{\infty} e^{-nh\nu/kT}\right)^3$$

$$= (1 + x + x^2 + x^3 + \cdots)^3$$

$$= 1 + 3(x + x^2 + x^3 + \cdots) + 3[x^2 + 2x(x^2 + x^3 + \cdots) + (x^2 + x^3 + \cdots)^2]$$

$$\quad + x^3 + 3x^2(x^2 + x^3 + \cdots) + 3x(x^2 + x^3 + \cdots)^2 + (x^2 + x^3 + \cdots)^3$$

$$= 1 + 3x + 6x^2 + 10x^3 + 15x^4 + 21x^5 + \cdots$$

$$= 1 + 3x + \frac{3.4}{2}x^2 + \frac{4.5}{2}x^3 + \frac{5.6}{2}x^4 + \frac{6.7}{2}x^5 + \cdots = \sum_{n=0}^{\infty} \frac{1}{2}(n + 1)(n + 2)x^n$$

$$= \sum_{n=0}^{\infty} \frac{1}{2}(n + 1)(n + 2)e^{-nh\nu/kT}. \quad (9.34)$$

Thus we obtain the equality[††]

$$I_N(3) = \left(e^{-h\nu/2kT} \sum_{n=0}^{\infty} e^{-nh\nu/kT}\right)^{3N} = I_{3N}(1). \quad (9.35)$$

[††]Perhaps this equality explains why in some literature this is considered obvious. See M.G. Bowler [6], p.37.

Example 9.4: The crystal treated as a system of phonons

Consider a crystal of N atoms as a system of phonons, *i.e.* vibrational excitations. Show that these phonons (1) obey Bose–Einstein statistics, (2) are not prescribed in number, so that $\epsilon_0 = 0$, (3) have energy values $\epsilon_i = h\nu_i$, (4) have statistical weight

$$\frac{d\mathcal{N}}{d\nu} = 4\pi V \left(\frac{1}{u_l^3} + \frac{2}{u_t^3} \right) \nu^2 \equiv aV\nu^2 \tag{9.36}$$

per unit frequency range. Calculate the thermodynamic functions for the crystal from the Bose–Einstein free energy of Eq. (7.26), *i.e.*

$$F = N\epsilon_0 + kT \sum_i g_i \ln[1 - e^{(\epsilon_0 - \epsilon_i)/kT}] \quad \text{with} \quad \epsilon_0 = 0. \tag{9.37}$$

Solution: The first part of the problem wants to underline the significance of phonons as the quanta of sound waves or atomic vibrational excitations in analogy to photons as the quanta of light waves. Points (1) to (4) have been dealt with in Secs. 9.1 and 9.2. The free energy to be considered is therefore

$$F = kT \sum_i g_i \ln[1 - e^{-h\nu_i/kT}], \tag{9.38}$$

which here becomes

$$F = kTaV \int_0^{\nu_{\max}} \nu^2 d\nu \ln[1 - e^{-h\nu/kT}]. \tag{9.39}$$

The entropy S follows from F with formula (5.22), *i.e.*

$$
\begin{aligned}
S &= -\left(\frac{\partial F}{\partial T} \right)_V = -kaV \int_0^{\nu_{\max}} \nu^2 d\nu \ln[1 - e^{-h\nu/kT}] \\
&\quad -kTaV \int_0^{\nu_{\max}} \nu^2 d\nu \frac{(-e^{-h\nu/kT})(h\nu/kT^2)}{1 - e^{-h\nu/kT}} \\
&= -kaV \int_0^{\nu_{\max}} \nu^2 d\nu \left[\ln\left(1 - e^{-h\nu/kT} \right) - \frac{h\nu}{kT} \left(\frac{1}{e^{h\nu/kT} - 1} \right) \right] \\
&\overset{(5.10)}{=} \frac{1}{T}(-F + E),
\end{aligned}
\tag{9.40}
$$

where

$$E = aV \int_0^{\nu_{\max}} \nu^2 d\nu \frac{h\nu}{e^{h\nu/kT} - 1}, \tag{9.41}$$

in agreement with Eq. (9.6).

Example 9.5: Using a momentum limit instead of frequency limit

Show that it would be more reasonable to take the lower wavelengths limits of longitudinal and transverse waves as equal, rather than the upper frequency limits as assumed by Debye. On this basis, show that it is best to reformulate the statistical theory in terms of momenta $p = h/\lambda = h\nu/u$. In terms of these momenta, evaluate the statistical weight per unit momentum range, and interpret the result.

Solution: We have $h\nu/kT = pu/kT$, $h\nu = pu$, $\nu = pu/h$, $d\nu = udp/h$. Then

$$d\mathcal{N} = 4\pi V \frac{3}{u^3} \nu^2 d\nu = 3\frac{4\pi V}{h^3} p^2 dp. \tag{9.42}$$

With these considerations one now has to impose an ultraviolet cutoff at $p_{\max}$ corresponding to a shortest length (the distance between atoms).

9.4 Problems without Worked Solutions

Example 9.6: Estimate of Einstein temperature

Consider a solid consisting of a cubic lattice with lattice constant a (shortest distance between neighbouring lattice points), and assume a spring-like interaction between neighbouring atoms at lattice points. Show that the compressibility κ of the solid and the spring constant κ_0 are related such that $\kappa \kappa_0 = a$. Taking one atom per lattice cube, express the lattice constant a in terms of atomic weight m_A, the density ρ of the material and Avogadro's number $N_A = 6.0 \times 10^{23}$. Then estimate roughly the Einstein temperature θ_E of Eq. (9.24) for such a solid made of copper (atomic weight $m_A = 63.5$, $\kappa = 4.5 \times 10^{-7} \text{bar}^{-1}$ (1 bar $= 10^6 \text{ g/cm s}^2$ or 10^5 newton /m^2) and density $\rho = 8.9 \text{ g cm}^{-3}$).
[Hint: κ is defined as $\kappa = [\rho(dP/d\rho)]^{-1}$, where P is the pressure, and κ_0 by the equation $m\ddot{x} + \kappa_0 x = 0$ to be compared with $m\ddot{x} + 4\pi^2 m\nu^2 x = 0$]. (Answer: $\theta_E = (\hbar/k)\sqrt{(N_A/\kappa m_A)(m_A/\rho N_A)^{1/3}}$.)[‡‡]

Example 9.7: Electron contribution to capacity of heat

In the case of some substances like O_2 and NO there are additional contributions to thermodynamical properties resulting from two closely lying electronic levels, say ϵ_1, ϵ_2 with $\epsilon_2 - \epsilon_1 \ll kT$, so that thermal excitation suffices in order to populate also the higher level. Establish the partition function for such a two-level system and calculate the electronic contribution C_{el} to the specific heat. What is the behaviour of C_{el} at high and at low temperatures T? What is the condition for a maximum of this quantity? For convenience set $\epsilon_1 = 0$ and $\epsilon = \epsilon_2 - \epsilon_1$. (Answer: The maximum of $C_{el}(T)$ is at temperature $T = 1/k\beta$ given by the solution of the equation $1 = (1/2)\epsilon\beta\tanh[\{\epsilon\beta + \ln(g_1/g_2)\}/2]$, where g_1, g_2 are the degeneracies of the two levels).

Example 9.8: Chemical potential derived from free energy

What is the free energy F of an Einstein solid of $3N$ one-dimensional oscillators, each of natural frequency ν, if the interaction between them lowers F by the amount AN, $A = \text{const.}$ From this derive the chemical potential ϵ_0 of the solid. What is the chemical potential of an ideal quantum gas of N particles in a volume V in the classical approximation?[§§]

Example 9.9: Bose–Einstein integrals and Riemann zeta functions $\zeta(x)$

Show that[¶¶]

$$\int_0^\infty \frac{x^2 e^{-x^2}}{1 - e^{-x^2}} dx = \frac{1}{4}\sqrt{\pi}\zeta\left(\frac{3}{2}\right), \quad \zeta\left(\frac{3}{2}\right) = 2.612. \tag{9.43}$$

More generally

$$\int_0^\infty \frac{x^n}{e^x - 1} dx = n!\,\zeta(n+1). \tag{9.44}$$

Thus Bose–Einstein integrals can be related to Riemann zeta functions.
[Hint: Use the change of variables in

$$\int_0^\infty x^2 dx\, e^{-x^2}[1 + \exp(-x^2) + \cdots] = \int_0^\infty x^2 dx \exp(-x^2)[1 + 2^{-3/2} + \cdots]].$$

Example 9.10: Entropy of free massless scalar fields

Show that the entropy S of a free scalar field in (a) $1+1$ dimensions (length L), and in (b) $1+3$ dimensions (volume V) is given by

$$(a)\; S = \frac{1}{3}\pi L T, \quad \text{and} \quad (b)\; S = \frac{2}{\pi^2}V\zeta(4)T^3, \tag{9.45}$$

[‡‡] F. Reif [59], Ex. 7.12.
[§§] From P.T. Landsberg [38], problem 14.9 on pp.272, 417.
[¶¶] R.B. Dingle [21], p.121.

where

$$\zeta(4) = \frac{1}{6} \int_0^\infty \frac{x^3 dx}{e^x - 1} = \frac{\pi^4}{90}. \tag{9.46}$$

[Hint: Use in (a) Eq. (6.23) and look at Eq. (9.6). Thus, with $\hbar = c = k = 1$ and with $(Tds)_V = dE$,

$$E = \int \frac{\epsilon_n dn}{e^{\epsilon_n/T} - 1} = \int \frac{2Ld\nu}{c} \frac{h\nu}{e^{h\nu/T} - 1} = \frac{1}{6}\pi LT^2, \quad \zeta(2) = \frac{\pi^2}{6}, \quad dS = \frac{L\pi}{3} dT].$$

Example 9.11: Debye frequency and velocity of sound
Assuming one ion in a cube of volume $(d/2)^3$ (d being the distance between neighbouring ions) and $u_t = u_l = u$, show that

$$\omega_D \simeq \frac{8u}{d}. \tag{9.47}$$

Example 9.12: Debye frequency in a superconductor
Considering the cubic metal lattice of a superconductor with neighbouring ion separation d at temperature $T = 0$, show that the Debye frequency is given by

$$\omega_D = \frac{2\sqrt{2}\hbar}{(m_e M a_B d^3)^{1/2}} \simeq \frac{2\sqrt{2}\hbar}{\sqrt{m_e M}\sqrt{n}d^2}, \tag{9.48}$$

where m_e is the mass of the electron, M that of an ion, a_B the Bohr radius, and $d \simeq n a_B$, n an integer.
[Hint: Equate the oscillator potential of the ion to the energy of attraction between electron and ion, and express e^2 in terms of a_B, giving $e^2 \simeq \hbar^2/m_e d$].

Example 9.13: Debye frequency evaluation
Evaluate the Debye frequency of Example 9.12 for mercury (atomic mass 0.331×10^{-24} kg) with $d = n$ Bohr radii $= n \times 0.529 \times 10^{-10}$ meters.
(Answer: $1.95 \times 10^{14}/n^{3/2} \sec^{-1}$).

Example 9.14: Equation of state
What is the equation of state of a gas of noninteracting bosonic particles?

Chapter 10

Electrons in Metals

10.1   Introductory Remarks

Electrons in metals present an example of *Fermi–Dirac statistics*. It is not correct to consider the electrons in a metal as an electron gas because this would have to be too highly compressed. Previously we had N = the number of atoms, here we have N = the number of electrons. **IF** (note the bigg if!) the electrons behaved classically, there would be a specific heat $3Nk/2$ (from 3 directions of motion per electron, no potential energy, and the internal energy E therefore classically given by $E = 3N \times kT/2$, and the specific heat C is obtained from this with $C = \partial E/\partial T$). This is quite large and is about $1/2$ of the lattice specific heat (*cf. e.g.* Eq. (9.16)) at high temperatures. This, however, is not observed (*i.e.* this specific heat $3Nk/2$). Hence the classical theory is no good. The correct theory is that using Fermi–Dirac statistics.

Why do electrons obey Fermi–Dirac statistics? Three reasons are:

1. Pauli's exclusion principle says: No two electrons in any one atom can have the same set of quantum numbers. Therefore Fermi–Dirac statistics applies here.

2. The electron spin is $1/2$. Therefore Fermi–Dirac statistics applies.

3. The electron is a real elementary particle with an elementary charge (it is not a compound particle). Therefore Fermi–Dirac statistics applies.

Hence the electrons must have a distribution function such that[*]

$$n_i = \frac{g_i}{e^{(\epsilon_i - \epsilon_0)/kT} + 1}. \tag{10.1}$$

---

[*]Throughout the entire text we use the symbol ϵ_0 with different meanings (Fermi energy, ground state energy, chemical potential); we assume the relevant meaning is clear from the respective context.

Here ϵ_0, called *Fermi energy*, is an important factor because the electrons are *conserved*.

It is evident that a realistic treatment of electrons in metals requires explicit use of the Fermi–Dirac distribution. In the following we begin with first and second order evaluations of the distribution for the calculation of the Fermi energy ϵ_0, the total energy E of a system of N electrons, and their specific heat C. The low-temperature behaviour of the Fermi–Dirac distribution, of ϵ_0, of the total energy E, and of the specific heat C are considered in detail in Examples 10.3 and 10.4. Similarly we consider in other examples the paramagnetism of a metal, and the thermionic emission of electrons from a metal. Finally we learn here why (so to speak) a table resists knocks by the knuckles of one's fist: Because of the Pauli principle! This principle applies also to baryons, the heavy fermions. Conservation of baryon number is the basis for the stability of matter.

10.2   Evaluation of the Distribution Function

We begin with the determination of ϵ_0. The energy ϵ_0 is obtained from the relation:

$$N = \sum_i n_i = \sum_i \frac{g_i}{e^{(\epsilon_i - \epsilon_0)/kT} + 1}. \tag{10.2}$$

What is the value of g_i, the *a priori* rating? It is

$$g_i = 2 \times \frac{4\pi V p^2 dp}{h^3}. \tag{10.3}$$

We have to multiply the usual value by 2, because there are two possible orientations of the electron spin. Now, nonrelativistically,

$$p^2 = 2m\epsilon, \quad \therefore \ 2pdp = 2md\epsilon, \quad \therefore \ pdp = md\epsilon. \tag{10.4}$$

Therefore:

$$g_i = \frac{8\pi V \sqrt{2m\epsilon}\, m d\epsilon}{h^3} = \frac{8\pi\sqrt{2}m^{3/2}V}{h^3}\epsilon^{1/2}d\epsilon \equiv g(\epsilon)d\epsilon. \tag{10.5}$$

It follows that

$$N = \sum_i n_i = \frac{8\pi\sqrt{2}m^{3/2}V}{h^3} \int_{0,\text{kinetic energy}}^{\infty} \frac{\epsilon^{1/2}d\epsilon}{e^{(\epsilon - \epsilon_0)/kT} + 1}. \tag{10.6}$$

This is the equation determining ϵ_0. The integral is called a *Fermi–Dirac integral*. We set $\epsilon_0 = kT_0$. Then T_0 comes out to be of the order of $50\,000°C$

for a typical metal. In the expression $(\epsilon - \epsilon_0)/kT$ the contribution ϵ_0/kT is always very large. When $\epsilon < \epsilon_0$, then $\epsilon - \epsilon_0 < 0$ and very large; when $\epsilon > \epsilon_0$, then $\epsilon - \epsilon_0 > 0$ and huge (simple consequence of $T \ll T_0$). We write

$$\frac{n_i}{g_i} = \frac{1}{e^{(\epsilon_i - \epsilon_0)/kT} + 1} \equiv f(\epsilon_i). \tag{10.7}$$

10.2.1 First approximation

We consider a first approximation for $\epsilon < \epsilon_0$. This is, since ϵ_0 is huge,

$$\frac{1}{e^{-\infty} + 1} = 1.$$

When $\epsilon \gg \epsilon_0$,

$$\frac{1}{e^{+\infty} + 1} = 0.$$

Hence the graph is the dashed line in Fig. 10.1.

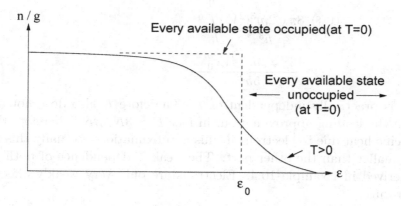

Fig. 10.1 The dashed line is the first approximation of n/g.

Strictly speaking this is exact only at absolute zero, and hence at high temperatures the edge is blurred as indicated by the continuous line in the graph of Fig. 10.1. To a first approximation any function which is integrated over the distribution function implies the relation

$$\int_0^\infty \frac{\phi(\epsilon)d\epsilon}{e^{(\epsilon - \epsilon_0)/kT} + 1} = \int_0^{\epsilon_0} \phi(\epsilon)d\epsilon. \tag{10.8}$$

Now we use this in the above evaluation of ϵ_0 from Eq. (10.6). Then to a first approximation:

$$N \stackrel{(10.8)}{=} \frac{8\pi\sqrt{2}m^{3/2}V}{h^3} \int_0^{\epsilon_0} \epsilon^{1/2}d\epsilon = \frac{2}{3}\frac{8\pi\sqrt{2}m^{3/2}V}{h^3}\epsilon_0^{3/2}. \tag{10.9}$$

This equation gives ϵ_0 in the dominant temperature-independent approximation, *i.e.*

$$\epsilon_0^{3/2} = \frac{3h^3}{16\sqrt{2}\pi m^{3/2}V}N, \qquad \epsilon_0 = \frac{h^2}{2m}\left(\frac{3N}{8\pi V}\right)^{2/3}. \tag{10.10}$$

We see that ϵ_0 is in this approximation independent of temperature. In
Example 10.3 we derive the next and temperature-dependent term of the
low-energy expansion of the Fermi energy ϵ_0. The approximation (10.10) of
ϵ_0 is conveniently set equal to kT_0. The temperature T_0 thus defined is called
degeneracy temperature, *i.e.* $(\epsilon_0)_{T=0} \equiv kT_0$.[†]
   Next we consider the total energy E in this approximation. We have

$$E = \sum_i n_i\epsilon_i$$

$$\stackrel{(10.6)}{=} \frac{8\pi\sqrt{2}m^{3/2}V}{h^3}\int_0^\infty \frac{\epsilon.\epsilon^{1/2}d\epsilon}{e^{(\epsilon-\epsilon_0)/kT}+1} \quad \text{(exact)}$$

$$\stackrel{(10.8)}{\simeq} \frac{8\pi\sqrt{2}m^{3/2}V}{h^3}\int_0^{\epsilon_0} \epsilon^{3/2}d\epsilon$$

$$\simeq \frac{16\pi\sqrt{2}m^{3/2}V\epsilon_0^{5/2}}{5h^3}. \tag{10.11}$$

But ϵ_0 is practically independent of T. Therefore E also does not depend
on T in the leading approximation, in fact $E \simeq 3N\epsilon_0/5$. Therefore there is
no specific heat due to electrons in this approximation. Actually this is very
much smaller than the other part. The weak T-dependence of both E and
C is derived in Example 10.4. Electrons are only very weakly affected by
temperature.

10.2.2   Second degree of approximation

For the second degree of approximation we expand the function $\phi(\epsilon)$ around
ϵ_0, *i.e.* we use the approximation

$$\phi(\epsilon) \simeq \phi(\epsilon_0) + (\epsilon - \epsilon_0)\phi'(\epsilon_0), \qquad \phi'(\epsilon_0) = \left(\frac{\partial\phi}{\partial\epsilon}\right)_{\epsilon=\epsilon_0}.$$

Then, as shown in Example 10.1, Eq. (10.35),

$$\int_0^\infty \frac{\phi(\epsilon)d\epsilon}{e^{(\epsilon-\epsilon_0)/kT}+1} \simeq \int_0^{\epsilon_0}\phi(\epsilon)d\epsilon + \frac{1}{6}(\pi kT)^2\phi'(\epsilon_0). \tag{10.12}$$

[†]See W. Shockley [70], p.242 or Ya.P. Terletskii [75], p.202.

The second term is important only because it is strongly temperature dependent. Therefore, from the original unapproximated Eq. (10.6),

$$
\frac{N}{8\pi\sqrt{2}m^{3/2}V/h^3} = \int_0^\infty \frac{\epsilon^{1/2}d\epsilon}{e^{(\epsilon-\epsilon_0)/kT}+1}
$$

$$
\overset{(10.12)}{=} \frac{2}{3}\epsilon_0^{3/2} + \frac{1}{6}(\pi kT)^2\frac{1}{2\epsilon_0^{1/2}}. \tag{10.13}
$$

But N is constant and independent of temperature. Therefore

$$
\frac{\partial}{\partial T}(\text{left hand side}) = 0.
$$

Hence (do not differentiate this)

$$
0 = \epsilon_0^{1/2}\frac{\partial\epsilon_0}{\partial T} + \frac{1}{6}\pi^2k^2T\epsilon_0^{-1/2} - O(\epsilon_0^{-3/2}), \tag{10.14}
$$

and hence

$$
\frac{\partial\epsilon_0}{\partial T} \simeq -\frac{1}{6}\pi^2k^2T\epsilon_0^{-1} + O(\epsilon_0^{-2}). \tag{10.15}
$$

We now repeat the same procedure for the energy

$$
E = \sum_i n_i\epsilon_i \overset{(10.2)}{=} \sum_i \frac{g_i\epsilon_i}{e^{(\epsilon_i-\epsilon_0)/kT}+1}. \tag{10.16}
$$

Then (effectively with an additional factor ϵ in the integrand of Eq. (10.13))

$$
\frac{E}{8\pi\sqrt{2}m^{3/2}V/h^3} \overset{(10.11)}{=} \int_0^\infty \frac{\epsilon\cdot\epsilon^{1/2}d\epsilon}{e^{(\epsilon-\epsilon_0)/kT}+1}
$$

$$
\overset{(10.12)}{=} \int_0^{\epsilon_0} \epsilon^{3/2}d\epsilon + \frac{1}{6}(\pi kT)^2\frac{3}{2}\epsilon_0^{1/2}
$$

$$
= \frac{2}{5}\epsilon_0^{5/2} + \frac{1}{6}(\pi kT)^2\frac{3}{2}\epsilon_0^{1/2}. \tag{10.17}
$$

Writing the *thermal capacity* (or specific heat) C of the N electrons in the system:

$$
C \simeq \frac{\partial E}{\partial T}, \tag{10.18}
$$

we get

$$
\frac{C}{8\pi\sqrt{2}m^{3/2}V/h^3} \simeq \epsilon_0^{3/2}\frac{\partial\epsilon_0}{\partial T} + \frac{1}{2}\pi^2k^2T\epsilon_0^{1/2}. \tag{10.19}
$$

Then (note both contributions on the right hand side are important for T small and ϵ_0 large)

$$\frac{C}{8\pi\sqrt{2}m^{3/2}V/h^3} \overset{(10.15)}{\simeq} -\frac{\pi^2k^2T}{6}\epsilon_0^{1/2} + \frac{1}{2}\pi^2k^2T\epsilon_0^{1/2}$$

$$= \frac{1}{3}\pi^2k^2T\epsilon_0^{1/2}. \tag{10.20}$$

We had[‡]

$$N \overset{(10.9)}{\simeq} \frac{16\pi\sqrt{2}m^{3/2}V}{3h^3}\epsilon_0^{3/2}, \tag{10.21}$$

or

$$\frac{N}{8\pi\sqrt{2}m^{3/2}V/h^3} = \frac{2}{3}\epsilon_0^{3/2}. \tag{10.22}$$

Dividing both sides of Eq. (10.20) by this expression, we obtain

$$\frac{C}{N} = \frac{1}{2}\pi^2k^2\frac{T}{\epsilon_0}, \tag{10.23}$$

or

$$C = \frac{1}{2}N\pi^2k\frac{T}{T_0}, \quad \text{since } (\epsilon_0)_{T=0} = kT_0. \tag{10.24}$$

This result shows that the specific heat of an electron gas is *of the order of*

$$\frac{T}{T_0} \times \text{ the classical value,}$$

(the classical value as noted in Sec. 10.1 being $3Nk/2$). But T/T_0 is very small since $T_0 \simeq 50\,000°$ absolute, so that this specific heat is very small (for further discussion see *e.g.* Terletskii [75], p.202. It is important only at very low temperatures (Debye) and at very high temperatures (where the lattice specific heat flattens out). Further aspects of electrons in metals are dealt with in Examples 10.5 to 10.9.[§]

10.3   Applications and Examples

Example 10.1: Evaluation of Fermi–Dirac integrals
Evaluate the Fermi–Dirac integral

$$I_{FD} = \int_0^\infty \frac{\phi(\epsilon)d\epsilon}{e^{(\epsilon-\epsilon_0)/kT} + 1}. \tag{10.25}$$

[‡]For discussion of this formula in the context of the band theory of metals see W. Shockley [70], pp.220-223.

[§]For additional discussion see *e.g.* S.K. Ma [42], Chapter 4 (on electrons in metals).

Solution: The following evaluation of the integral is that of Dingle.[¶] We set:

$$\Phi(\epsilon) = \int_0^\epsilon \phi(\epsilon)d\epsilon, \quad \Phi'(\epsilon) = \phi(\epsilon). \tag{10.26}$$

Integrating the integral I_{FD} by parts we obtain

$$\int_0^\infty \frac{\phi(\epsilon)d\epsilon}{e^{(\epsilon-\epsilon_0)/kT}+1} = \left[\frac{\Phi(\epsilon)}{e^{(\epsilon-\epsilon_0)/kT}+1}\right]_0^\infty + \frac{1}{kT}\int_0^\infty \frac{\Phi(\epsilon)e^{(\epsilon-\epsilon_0)/kT}d\epsilon}{[e^{(\epsilon-\epsilon_0)/kT}+1]^2}. \tag{10.27}$$

The first term vanishes at $\epsilon = \infty$ owing to the exponential factor in the denominator, and at $\epsilon = 0$ owing to the vanishing of $\Phi(\epsilon)$ there. When $\epsilon_0 \gg kT$ (*i.e.* at low temperatures), the factor

$$\frac{e^{(\epsilon-\epsilon_0)/kT}}{[e^{(\epsilon-\epsilon_0)/kT}+1]^2}$$

appearing in the integral of the second term is negligible except when $\epsilon \simeq \epsilon_0$. Expanding $\Phi(\epsilon)$ about this point, we have

$$\Phi(\epsilon) = \Phi(\epsilon_0) + (\epsilon-\epsilon_0)\Phi'(\epsilon_0) + \frac{1}{2}(\epsilon-\epsilon_0)^2\Phi''(\epsilon_0) + \cdots. \tag{10.28}$$

We set

$$x = \frac{\epsilon-\epsilon_0}{kT}, \quad d\epsilon = kTdx,$$

and replace the lower limit $\epsilon = 0$ or $x = -\epsilon_0/kT$ of the integral by $-\infty$. Since the integrand is negligible there anyway, we obtain

$$\int_0^\infty \frac{\phi(\epsilon)d\epsilon}{e^{(\epsilon-\epsilon_0)/kT}+1} = \int_{x=-\epsilon_0/kT}^\infty \frac{\Phi(\epsilon)e^x}{(e^x+1)^2}dx$$

$$\overset{(10.28)}{\simeq} \Phi(\epsilon_0)\int_{-\infty}^\infty \frac{e^x dx}{(e^x+1)^2} + kT\Phi'(\epsilon_0)\int_{-\infty}^\infty \frac{xe^x dx}{(e^x+1)^2}$$

$$+ (kT)^2\Phi''(\epsilon_0)\frac{1}{2}\int_{-\infty}^\infty \frac{x^2e^x dx}{(e^x+1)^2} + \cdots. \tag{10.29}$$

Now, considering the first integral,

$$\int_{-\infty}^\infty \frac{e^x dx}{(e^x+1)^2} = \int_{e^x=0}^\infty \frac{d(e^x)}{(e^x+1)^2} = -\left[\frac{1}{e^x+1}\right]_{e^x=0}^\infty = 1. \tag{10.30}$$

Also, in the case of the second integral,

$$\int_{-\infty}^\infty \frac{xe^x dx}{(e^x+1)^2} = 0, \tag{10.31}$$

since here the integrand is odd in x, *i.e.*

$$\frac{x}{(e^{x/2}+e^{-x/2})^2}.$$

[¶]R.B. Dingle, *Asymptotic Expansions for Integrals of Fermi–Dirac Statistics*, notes supplied as addendum to problem sheets, University of W.A. (1956). For further consideration of Fermi–Dirac integrals see R.B. Dingle [17], pp.38, 109-110, 436, and R.B. Dingle [18]. For formulas and Tables see also the chapter in the NIST Handbook [53]. An early reference is that of A. Sommerfeld [71].

The third integral has an even integrand and can be evaluated as follows:

$$\frac{1}{2} \int_{-\infty}^{\infty} \frac{x^2 e^x \, dx}{(e^x + 1)^2} = \int_0^{\infty} \frac{x^2 e^x \, dx}{(e^x + 1)^2} = \int_0^{\infty} \frac{x^2 e^{-x} \, dx}{(1 + e^{-x})^2}$$

$$= \int_0^{\infty} x^2 \, dx \left(e^{-x} - 2e^{-2x} + 3e^{-3x} - \cdots \right). \tag{10.32}$$

Since *e.g.*

$$\int_0^{\infty} x^2 e^{-2x} \, dx = \frac{1}{2^3} \int_0^{\infty} (2x)^2 e^{-(2x)} \, d(2x), \tag{10.33}$$

the integrals appearing in the second and higher order terms can be expressed as multiples of the leading integral, so that

$$\frac{1}{2} \int_{-\infty}^{\infty} \frac{x^2 e^x \, dx}{(e^x + 1)^2} = \int_0^{\infty} x^2 e^{-x} \, dx \left(1 - \frac{2}{2^3} + \frac{3}{3^3} - \cdots \right)$$

$$= 2 \left(1 - \frac{1}{2^2} + \frac{1}{3^2} - \cdots \right) = 2 \left(\frac{\pi^2}{12} \right) = \frac{\pi^2}{6}, \tag{10.34}$$

where we used the factorial integral (12.151) and in the last step one of the known expansions for π.[||] Remembering that $\Phi(\epsilon)$ was defined by Eq. (10.26), we obtain finally

$$\int_0^{\infty} \frac{\phi(\epsilon) d\epsilon}{e^{(\epsilon - \epsilon_0)/kT} + 1} = \Phi(\epsilon_0) + (kT)^2 \Phi''(\epsilon_0) \frac{\pi^2}{6} + \cdots$$

$$= \int_0^{\epsilon_0} \phi(\epsilon) d\epsilon + \frac{1}{6} (\pi kT)^2 \phi'(\epsilon_0) + \cdots. \tag{10.35}$$

Example 10.2: Low temperature behaviour of the distribution function

Show that at the absolute zero of temperature, the Fermi–Dirac distribution function

$$n_i = \frac{g_i}{e^{(\epsilon_i - \epsilon_0)/kT} + 1} \equiv g_i f(\epsilon_i)$$

means that all cells are fully occupied if they correspond to energies less than ϵ_0, but are quite empty if they correspond to energies greater than ϵ_0. Replace the discrete energies ϵ_i by a continuum by putting

$$f(\epsilon) = \frac{1}{e^{(\epsilon - \epsilon_0)/kT} + 1}, \quad g_i = g(\epsilon) d\epsilon, \tag{10.36}$$

and show that under most conditions

$$\int f(\epsilon) g(\epsilon) d\epsilon = - \int G(\epsilon) f'(\epsilon) d\epsilon, \quad \text{where } G(\epsilon) = \int_0^{\epsilon} g(\epsilon) d\epsilon. \tag{10.37}$$

Show that at low temperatures ($kT \ll \epsilon_0$), $f(\epsilon)$ only varies near $\epsilon = \epsilon_0$, and hence that

$$\int f(\epsilon) g(\epsilon) d\epsilon \simeq G(\epsilon_0) - G'(\epsilon_0) \int_{\epsilon - \epsilon_0 = -\infty}^{\infty} (\epsilon - \epsilon_0) f'(\epsilon) d\epsilon$$

$$- \frac{1}{2} G''(\epsilon_0) \int_{\epsilon - \epsilon_0 = -\infty}^{\infty} (\epsilon - \epsilon_0)^2 f'(\epsilon) d\epsilon$$

$$= G(\epsilon_0) + \frac{1}{6} (\pi kT)^2 G''(\epsilon_0) = \int_0^{\epsilon_0} g(\epsilon) d\epsilon + \frac{1}{6} (\pi kT)^2 g'(\epsilon_0). \tag{10.38}$$

[||] H.B. Dwight [23], formula 48.3, p.11.

Estimate the errors introduced by these approximations.

Solution: The derivation of the result proceeds parallel to the derivation of the result of Example 10.1 with (there) ϕ replaced by (here) g and Φ by G. The errors introduced in this derivation are of two types. The error of the first approximation, *i.e.* of replacing $-\epsilon_0/kT$ by $-\infty$, implies on the right hand side of Eq. (10.29) an additive error of

$$-\int_{-\infty}^{-\epsilon_0/kT} \Phi(\epsilon) \frac{e^x\,dx}{(e^x+1)^2} \simeq -\int_{-\infty}^{-\epsilon_0/kT} \Phi(\epsilon)e^x\,dx \simeq \Phi(\epsilon_0)\int_{-\infty}^{-\epsilon_0/kT} e^x\,dx$$

$$\simeq -\Phi(\epsilon_0)e^{-\epsilon_0/kT}. \tag{10.39}$$

The error of the other approximation of truncating the series with the quadratic term is of $O[(\pi kT)^3]$.

Example 10.3: Fermi energy ϵ_0 as a function of T

Noting that the number of electrons in a metal of volume V would be (*cf.* Eq. (10.6))

$$N = \sum_i n_i = \int f(\epsilon)g(\epsilon)d\epsilon = \frac{4\pi V}{h^3}(2m)^{3/2}\int_0^\infty f(\epsilon)\sqrt{\epsilon}d\epsilon, \quad g(\epsilon) = \frac{1}{h^3}8\pi\sqrt{2}m^{3/2}V\epsilon^{1/2}, \tag{10.40}$$

(the extra factor 2 compared with the classical expression coming about because of the 2 possible spin orientations of an electron), use the formula of Example 10.2 to calculate ϵ_0 as a function of T.

Solution: From Example 10.2 we obtain:

$$\int f(\epsilon)g(\epsilon)d\epsilon = \int_0^{\epsilon_0} g(\epsilon)d\epsilon + \frac{1}{6}(\pi kT)^2 g'(\epsilon_0)$$

$$= \frac{8\pi\sqrt{2}m^{3/2}V}{h^3}\left[\int_0^{\epsilon_0}\epsilon^{1/2}d\epsilon + \frac{1}{6}(\pi kT)^2\frac{1}{2\sqrt{\epsilon_0}}\right]$$

$$= \frac{8\pi\sqrt{2}m^{3/2}V}{h^3}\left[\frac{2}{3}\epsilon_0^{3/2} + \frac{1}{12\sqrt{\epsilon_0}}(\pi kT)^2\right]. \tag{10.41}$$

From Eq. (10.40) we obtain therefore

$$\frac{Nh^3}{4\pi V}\frac{1}{(2m)^{3/2}} = \frac{2}{3}\epsilon_0^{3/2} + \frac{1}{12\sqrt{\epsilon_0}}(\pi kT)^2, \tag{10.42}$$

which is also Eq. (10.13). In order to solve the equation for ϵ_0 we set

$$a := \frac{Nh^3}{4\pi V(2m)^{3/2}}, \quad b := \frac{2}{3}, \quad c := \frac{1}{12}(\pi kT)^2. \tag{10.43}$$

Then

$$\epsilon_0^{3/2} = \frac{a}{b} - \frac{c}{b\epsilon_0^{1/2}} \simeq \frac{a}{b} - \frac{c}{b(a/b)^{1/3}} = \frac{a}{b}\left[1 - \frac{cb^{1/3}}{a^{4/3}}\right], \tag{10.44}$$

and (in the second step remember that $b = 2/3$)

$$\epsilon_0 = \left(\frac{a}{b}\right)^{2/3}\left[1 - \frac{cb^{1/3}}{a^{4/3}}\right]^{2/3} \simeq \left(\frac{a}{b}\right)^{2/3}\left[1 - \left(\frac{b}{a}\right)^{4/3}c\right], \tag{10.45}$$

where

$$\left(\frac{b}{a}\right)^{1/3} = \frac{2\sqrt{2m}}{h}\left(\frac{\pi V}{3N}\right)^{1/3}. \tag{10.46}$$

Owing to the factor $c = (\pi kT)^2/12$ the result (10.45) now exhibits a weak temperature dependence of the Fermi energy ϵ_0.[**]

For later reference we note here that with the help of Eqs. (10.10) and (10.43):

$$
\begin{aligned}
\frac{1}{a}\sqrt{\epsilon_0} &= \frac{4\pi V(2m)^{3/2}}{Nh^3}\left(\frac{h^2}{2m}\right)^{1/2}\left(\frac{3N}{8\pi V}\right)^{1/3} = \frac{2m}{h^2}\left(\frac{4\pi V}{N}\right)\left(\frac{3N}{8\pi V}\right)^{1/3} \\
&= 3\frac{2m}{h^2}\left(\frac{4\pi V}{3N}\right)\left(\frac{3N}{4\pi V}\right)^{1/3}\frac{1}{2^{1/3}} = 3\frac{2m}{h^2}\left(\frac{4\pi V}{3N}\right)^{2/3}\frac{1}{2^{1/3}} \\
&= 3\frac{2m}{2^{1/3}}\left(\frac{4\pi V}{3Nh^3}\right)^{2/3}.
\end{aligned}
\tag{10.47}
$$

Thus — since the quantity a defined in Eq. (10.43) involves only the ratio N/V — the Fermi energy ϵ_0 also depends only on the ratio N/V, *i.e.* the density. Equation (10.47) will be required later in the derivation of the result (10.72).

Example 10.4: Energy and specific heat of electrons in metals

Noting that (with $f(\epsilon)$ given by Eq. (10.36))

$$
E \stackrel{(10.11)}{=} \frac{4\pi V}{h^3}(2m)^{3/2}\int_0^\infty f(\epsilon)\epsilon^{3/2}d\epsilon,
\tag{10.48}
$$

calculate E as a function of T and $\epsilon_0(T)$, and hence by Example 10.3 of T. Then show that the specific heat C of the system is

$$
C = \left(\frac{\partial E}{\partial T}\right) = \frac{1}{2}\pi^2 Nk\frac{T}{T_0},
\tag{10.49}
$$

where T_0 is the *degeneracy temperature* given by

$$
kT_0 = (\epsilon_0)_{T=0} = \frac{1}{2m}\left(\frac{3\hbar^3\pi^2 N}{V}\right)^{2/3}.
\tag{10.50}
$$

Solution: Proceeding as in Example 10.3 we have in the present case (with Eqs. (10.45), (10.43))

$$
\begin{aligned}
E \stackrel{(10.40)}{=} \frac{8\pi\sqrt{2}m^{3/2}V}{h^3}\int_0^\infty f(\epsilon)\epsilon^{3/2}d\epsilon &\stackrel{(10.38)}{=} \frac{8\pi\sqrt{2}m^{3/2}V}{h^3}\left[\int_0^{\epsilon_0}\epsilon^{3/2}d\epsilon + \frac{1}{6}(\pi kT)^2\frac{3}{2}\epsilon_0^{1/2}\right] \\
&= \frac{8\pi\sqrt{2}m^{3/2}V}{h^3}\left[\frac{2}{5}\epsilon_0^{5/2} + \frac{1}{4}(\pi kT)^2\epsilon_0^{1/2}\right] = \frac{8\pi\sqrt{2}m^{3/2}V}{h^3}\frac{2}{5}\epsilon_0^{5/2}\left[1 + \frac{5}{8}\frac{(\pi kT)^2}{\epsilon_0^2}\right] \\
&\simeq \frac{8\pi\sqrt{2}m^{3/2}V}{h^3}\frac{2}{5}\epsilon_0^{5/2}\left[1 + \frac{15}{2}\left(\frac{b}{a}\right)^{4/3}c\right].
\end{aligned}
\tag{10.51}
$$

Since $\epsilon_0 \propto (V/N)^{4/3}$, this implies $E \propto V$ if N/V is constant. Substituting for ϵ_0 we obtain:

$$
\begin{aligned}
E \stackrel{(10.45)}{=} \frac{8\pi\sqrt{2}m^{3/2}V}{h^3}\frac{2}{5}\left(\frac{a}{b}\right)^{5/3}\left[1 - \left(\frac{b}{a}\right)^{4/3}c\right]^{5/2}\left[1 + \frac{15}{2}\left(\frac{b}{a}\right)^{4/3}c\right] \\
\simeq \frac{8\pi\sqrt{2}m^{3/2}V}{h^3}\frac{2}{5}\left(\frac{a}{b}\right)^{5/3}\left[1 - \frac{5}{2}\left(\frac{b}{a}\right)^{4/3}c + \frac{15}{2}\left(\frac{b}{a}\right)^{4/3}c\right] \\
= \frac{16\pi\sqrt{2}m^{3/2}V}{5h^3}\left(\frac{a}{b}\right)^{5/3}\left[1 + 5\left(\frac{b}{a}\right)^{4/3}c\right].
\end{aligned}
\tag{10.52}
$$

[**]Similar and further considerations may be found for instance in P.T. Landsberg [39], pp.275-279.

Here the factor $c = (\pi k T)^2/12$ provides the temperature dependence of the energy of the electrons. The dominant factor agrees with the T-independent approximation (10.11) (*cf.* Eq. (10.44)). The specific heat C of the electrons in the metal is now given by

$$
\begin{aligned}
C = \frac{\partial E}{\partial T} &= \frac{16\pi\sqrt{2}m^{3/2}V}{5h^3} 5 \left(\frac{a}{b}\right)^{1/3} \frac{\partial c}{\partial T} = \frac{16\pi^3\sqrt{2}m^{3/2}Vk^2T}{6h^3} \left(\frac{a}{b}\right)^{1/3} \\
&\overset{(10.46)}{=} \frac{8\pi^3\sqrt{2}m^{3/2}Vk^2T}{3h^3} \frac{(3N)^{1/3}}{2\sqrt{2m}(\pi V/h^3)^{1/3}} \\
&= \frac{4\pi^3 mk^2T}{3(h^3/V)} \left(\frac{3Nh^3}{\pi V}\right)^{1/3} = \frac{4\pi^2 mk^2T}{3(h^3/V)} \left(\frac{3Nh^3\pi^2}{V}\right)^{1/3} \\
&= \frac{4N\pi^4 mk^2T}{(3Nh^3\pi^2/V)^{2/3}} \overset{h\to\hbar}{=} \frac{4N\pi^4 mk^2T}{(3N\pi^2\hbar^3/V)^{2/3}(2\pi)^2} \\
&= \frac{1}{2}\pi^2 NkT \frac{2mk}{(3N\pi^2\hbar^3/V)^{2/3}}.
\end{aligned}
\tag{10.53}
$$

This result can be rewritten in the form of Eq. (10.24).

Example 10.5: The free energy of electrons in a metal

Show that the free energy on Fermi–Dirac statistics is (with $g(\epsilon)$ as in Eq. (10.40))

$$
F \simeq N\epsilon_0 - kT \int_0^\infty \ln[1 + e^{(\epsilon_0 - \epsilon)/kT}] g(\epsilon) d\epsilon, \quad g(\epsilon) = \frac{1}{h^3} 8\pi\sqrt{2}m^{3/2}V\epsilon^{1/2}.
\tag{10.54}
$$

Cast the integral into the form of Example 10.2 by integration by parts, and hence calculate the free energy of N electrons as a function of T and $\epsilon_0(T)$, and hence by Example 10.3 of T. Noting that $\partial F/\partial\epsilon_0 = 0$, determine ϵ_0, S, E and C as functions of T.

Solution: In Eq. (7.26) we replace $\sum_i g_i$ by $\int g(\epsilon)d\epsilon$ and thus obtain Eq. (10.54) for Fermi–Dirac statistics. Performing the partial integration we obtain

$$
\begin{aligned}
&\int_0^\infty \ln[1 + e^{(\epsilon_0 - \epsilon)/kT}] g(\epsilon) d\epsilon \\
&= \left[\ln[1 + e^{(\epsilon_0 - \epsilon)/kT}] \underbrace{\int_0^\epsilon g(\epsilon)d\epsilon}_{G(\epsilon)} \right]_0^\infty + \frac{1}{kT} \int_0^\infty \frac{e^{(\epsilon_0 - \epsilon)/kT} G(\epsilon)d\epsilon}{(1 + e^{(\epsilon_0 - \epsilon)/kT})} \\
&= \frac{1}{kT} \int_0^\infty \frac{G(\epsilon)d\epsilon}{e^{(\epsilon - \epsilon_0)/kT} + 1} \\
&\overset{(10.35)}{=} \frac{1}{kT} \left[\int_0^{\epsilon_0} G(\epsilon)d\epsilon + \frac{1}{6}(\pi kT)^2 G'(\epsilon_0) + \cdots \right].
\end{aligned}
\tag{10.55}
$$

Hence the free energy is

$$
F = N\epsilon_0 - \int_0^{\epsilon_0} G(\epsilon)d\epsilon - \frac{1}{6}(\pi kT)^2 G'(\epsilon_0) + \cdots .
\tag{10.56}
$$

Here

$$
\begin{aligned}
G(\epsilon) &= \int_0^\epsilon g(\epsilon)d\epsilon = \int_0^\epsilon \frac{8\pi\sqrt{2}m^{3/2}V}{h^3}\epsilon^{1/2}d\epsilon = \frac{8\pi\sqrt{2}m^{3/2}V}{h^3}\frac{2}{3}\epsilon^{3/2}, \\
\int_0^{\epsilon_0} G(\epsilon)d\epsilon &= \frac{16\pi\sqrt{2}m^{3/2}V}{3h^3}\frac{2}{5}\epsilon_0^{5/2},
\end{aligned}
\tag{10.57}
$$

so that

$$G'(\epsilon) = g(\epsilon) = \frac{8\pi\sqrt{2}m^{3/2}V}{h^3}\epsilon^{1/2}. \tag{10.58}$$

It follows that the free energy becomes

$$F = N\epsilon_0 - \frac{16\pi\sqrt{2}m^{3/2}V}{3h^3}\frac{2}{5}\epsilon_0^{5/2} - \frac{(\pi kT)^2}{6}\frac{8\pi\sqrt{2}m^{3/2}V}{h^3}\epsilon_0^{1/2}, \tag{10.59}$$

and in terms of the parameters a, b, c defined by Eq. (10.43) this becomes

$$\frac{F}{N} = \epsilon_0 - \frac{4}{15a}\epsilon_0^{5/2} - \frac{2c}{a}\epsilon_0^{1/2}. \tag{10.60}$$

We recall from Eq. (7.27) the relation

$$\frac{\partial F}{\partial \epsilon_0} = 0.$$

Applying this relation to the free energy of Eq. (10.60), we obtain

$$0 = 1 - \frac{2}{3a}\epsilon_0^{3/2} - \frac{c}{a}\epsilon_0^{-1/2} \quad \text{or} \quad 0 = 1 - \frac{b}{a}\epsilon_0^{3/2} - \frac{c}{a}\epsilon_0^{-1/2}. \tag{10.61}$$

Multiplying this equation by a/b, we obtain

$$0 = \frac{a}{b} - \epsilon_0^{3/2} - \frac{c}{b}\epsilon_0^{-1/2}, \quad \text{or} \quad \epsilon_0^{3/2} = \frac{a}{b}\left[1 - \frac{c}{a\epsilon_0^{1/2}}\right], \tag{10.62}$$

which is Eq. (10.44) (unapproximated). The entropy S now follows from the thermodynamical relation $F = E - TS$ as a function of T. Similarly the energy E and the specific heat C are obtained as functions of T, and the results will obviously be identical with those obtained above.

Example 10.6: Pauli spin paramagnetism of a metal

Calculate the Pauli spin paramagnetism of a metal in a magnetic field H by taking (this is half of Eq. (10.54) for spins parallel to H, plus half with spins antiparallel to H)

$$F = N\epsilon_0 - \frac{2\pi V(2m)^{3/2}kT}{h^3}\left[\int \ln[1 + e^{(\epsilon_0 - \epsilon')/kT}]\sqrt{\epsilon}d\epsilon + \int \ln[1 + e^{(\epsilon_0 - \epsilon'')/kT}]\sqrt{\epsilon}d\epsilon\right], \tag{10.63}$$

where

$$\epsilon' = \frac{p^2}{2m} + \mu H \quad \text{(spins against field)}, \qquad \epsilon'' = \frac{p^2}{2m} - \mu H \quad \text{(spins with field)}, \tag{10.64}$$

with

$$\mu = \frac{e\hbar}{2mc} \qquad \text{being the Bohr magneton.} \tag{10.65}$$

Calculate the magnetic moment M from the relation

$$M = -\left(\frac{\partial F}{\partial H}\right)_{T,V} = -\left\{\left(\frac{\partial F}{\partial H}\right)_{T,V,\epsilon_0} + \frac{\partial F}{\partial \epsilon_0}\frac{d\epsilon_0}{dH}\right\} = -\left(\frac{\partial F}{\partial H}\right)_{T,V,\epsilon_0}. \tag{10.66}$$

Show that the value so obtained is approximately T/T_0 times that yielded by Maxwell–Boltzmann statistics, and explain the difference.

Solution: From Eq. (10.60) in Example 10.5 we obtain the free energy F, i.e. from

$$\frac{F}{N} = \epsilon_0 - \frac{4}{15a}\epsilon_0^{5/2} - \frac{2c}{a}\epsilon_0^{1/2}, \quad \epsilon_0 \gg \mu H. \tag{10.67}$$

With replacements $\epsilon_0 \to \epsilon_0 \pm \mu H$ in $F \to F_\pm$ we obtain:

$$\frac{F_\pm}{N} = (\epsilon_0 \pm \mu H) - \frac{4}{15a}(\epsilon_0 \pm \mu H)^{5/2} - \frac{2c}{a}(\epsilon_0 \pm \mu H)^{1/2}, \tag{10.68}$$

and by expanding in powers of $\mu H/\epsilon_0$:

$$\begin{aligned}
\frac{F_+}{N} &= (\epsilon_0 + \mu H) - \frac{4}{15a}\epsilon_0^{5/2}\left(1 + \frac{\mu H}{\epsilon_0}\right)^{5/2} - \frac{2c}{a}\epsilon_0^{1/2}\left(1 + \frac{\mu H}{\epsilon_0}\right)^{1/2} \\
&= \left\{\epsilon_0 - \frac{4}{15a}\epsilon_0^{5/2} - \frac{2c}{a}\epsilon_0^{1/2}\right\} + \mu H\left\{1 - \frac{2}{3a}\epsilon_0^{3/2} - \frac{c}{a}\epsilon_0^{-1/2}\right\} \\
&\quad - \frac{1}{2a}\mu^2 H^2\epsilon_0^{1/2} + \frac{c}{4a}\mu^2 H^2\epsilon_0^{-3/2} + \cdots.
\end{aligned} \tag{10.69}$$

Hence, differentiating with respect to H, we obtain

$$\frac{\partial}{\partial H}\left(\frac{F_+}{N}\right) = \mu\left\{1 - \frac{2\epsilon_0^{3/2}}{3a} - \frac{c}{a\epsilon_0^{1/2}}\right\} - \frac{\mu^2 H}{a}\sqrt{\epsilon_0} + \frac{c\mu^2 H}{2a\epsilon_0^{3/2}}. \tag{10.70}$$

It follows therefore that for $\epsilon_0 \gg \mu H$,

$$-\frac{M}{N} = \frac{1}{2}\frac{\partial}{\partial H}\left(\frac{F_+ + F_-}{N}\right) = \frac{1}{2}2\left(-\frac{\mu^2 H}{a}\sqrt{\epsilon_0}\right) = -\frac{\mu^2 H}{a}\sqrt{\epsilon_0}. \tag{10.71}$$

Thus with Eq. (10.47):

$$\begin{aligned}
M &= N\mu^2 H.3.\frac{2m}{2^{1/3}}\left(\frac{4\pi V}{3Nh^3}\right)^{2/3} = \frac{3\mu^2 N H 2^{4/3}}{2^{1/3}}\left(\frac{\pi V}{3Nh^3}\right)^{2/3}2m \\
&= 6\mu^2 NH\left(\frac{\pi V}{3\hbar^3 N 2^3\pi^3}\right)^{2/3}2m = \frac{6\mu^2 NH}{2^2}\underbrace{\left(\frac{V}{3\hbar^3 N\pi^2}\right)^{2/3}2m}_{1/kT_0}
\end{aligned}$$

$$\overset{(10.50)}{=} \frac{3}{2}\frac{\mu^2 NH}{kT}\left(\frac{kT}{kT_0}\right) = \frac{3}{2}\frac{\mu^2 NH}{kT}\left(\frac{T}{T_0}\right). \tag{10.72}$$

This result may now be compared with the classical Maxwell–Boltzmann expression (4.46), $\overline{M} = M^2 H/3kT$ per particle. Apart from the different statistics, the Maxwell–Boltzmann result cannot be valid near $T \sim 0$, whereas the present result taking the degeneracy into account is valid near $T \sim 0$.[††]

Example 10.7: Thermionic emission of electrons from a metal[‡‡]

The classical treatment of thermionic emission of electrons from a metal surface was considered in Example 2.8. Here the Fermi–Dirac theory is to be considered. Show that the number n of electrons which can escape from unit surface area in unit time is given by

$$n = \frac{4\pi mk^2 T^2}{h^3} F_1\left(\frac{\epsilon_0 - \epsilon_{\text{surface}}}{kT}\right), \tag{10.73}$$

where the function F_1 is a Fermi–Dirac integral defined by ($x = \epsilon/kT$)

$$F_\kappa(\eta) = \int_0^\infty \frac{x^\kappa}{e^{x-\eta} + 1} dx = \frac{1}{(kT)^{\kappa+1}} \int_0^\infty \frac{\epsilon^\kappa d\epsilon}{e^{(\epsilon-\epsilon_0)/kT} + 1}. \tag{10.74}$$

Solution: The *a priori* weighting per unit volume is in the present case, and with due regard to the two possible spin orientations of an electron,

$$\frac{2(mdu)(mdv)(mdw)}{h^3},$$

where u, v, w are the velocity components of the electron. Thus N, the number of electrons per unit volume in the metal is (the volume V in the weighting having been divided out),

$$N = \frac{2m^3}{h^3} \int \int \int \frac{dudvdw}{e^{(\epsilon-\epsilon_0)/kT} + 1}. \tag{10.75}$$

Here we can replace $dudvdw$ by $4\pi c^2 dc$, where ($\epsilon = mc^2/2$),

$$c^2 = u^2 + v^2 + w^2 = \frac{2\epsilon}{m}, \quad \text{so that} \quad dudvdw = 4\pi c^2 dc = \frac{4\pi\sqrt{2}}{m^{3/2}} \epsilon^{1/2} d\epsilon. \tag{10.76}$$

Thus

$$N = \frac{8\sqrt{2}\pi m^{3/2}}{h^3} \int_0^\infty \frac{\epsilon^{1/2} d\epsilon}{e^{(\epsilon-\epsilon_0)/kT} + 1} = \frac{8\sqrt{2}\pi (mkT)^{3/2}}{h^3} F_{1/2}\left(\frac{\epsilon_0}{kT}\right), \tag{10.77}$$

with the Fermi–Dirac integral defined as above.

Now considering thermionic emission, we wish to compute the number n of electrons which can escape from unit area in unit time. This means we have to distinguish between the velocity components u aiming at the surface, and components v and w orthogonal to this direction, and these have to be treated differently in the integral of Eq. (10.75). Thus for a clear distinction, we first note again that N, the number of electrons per unit volume in the metal is as in Eq. (10.75) given by

$$N = \frac{2m^3}{h^3} \int \int \int \frac{dudvdw}{e^{(\epsilon-\epsilon_0)/kT} + 1},$$

[††]For further discussion see W. Pauli [54], and N.F. Mott and H. Jones [52], pp.184-189.
[‡‡]The solutions of Examples 10.7 to 10.9 are based on notes of R.B. Dingle.

but the number of electrons which can escape in the u-direction from unit surface in unit time is

$$n = \frac{2m^3}{h^3} \int u\,du \int \int \frac{dv\,dw}{e^{(\epsilon-\epsilon_0)/kT} + 1} \tag{10.78}$$

($dN \times$ velocity = number per unit area in unit time = dn). It is now convenient to integrate here first over $dv\,dw$, since the emission condition is independent of v and w. Writing $c^2 = v^2 + w^2$, $dv\,dw = 2\pi c.dc = d(\pi c^2)$, we have

$$\mathcal{I} := \int_{-\infty}^{\infty} \int_{-\infty}^{\infty} \frac{dv\,dw}{\exp[\{m(u^2 + v^2 + w^2)/2 - \epsilon_0\}/kT] + 1} = \pi \int_0^{\infty} \frac{d(c^2)}{ae^{mc^2/2kT} + 1}, \tag{10.79}$$

where

$$a = e^{(mu^2/2 - \epsilon_0)/kT}. \tag{10.80}$$

and

$$n = \frac{2m^3}{h^3} \int u\,du\,\mathcal{I}. \tag{10.81}$$

The definite integral $\mathcal{I}$ can be evaluated with the help of the following indefinite integral (this can be verified by differentiation):

$$\int \frac{dt}{ae^{pt} + b} = -\frac{1}{bp} \ln\left(1 + \frac{b}{a}e^{-pt}\right). \tag{10.82}$$

With this formula the integral $\mathcal{I}$ becomes:

$$\begin{aligned}
\mathcal{I} &= \pi\left(-\frac{1}{m/2kT}\right) \ln\left[1 + e^{-(mu^2/2-\epsilon_0)/kT} e^{-mc^2/2kT}\right]\Big|_{c^2=0}^{c^2=\infty} \\
&= -\frac{2\pi kT}{m}\left[\ln(1) - \ln\left\{1 + e^{-(mu^2/2-\epsilon_0)/kT}\right\}\right] \\
&= \frac{2\pi kT}{m} \ln\left[1 + e^{(\epsilon_0 - mu^2/2)/kT}\right].
\end{aligned} \tag{10.83}$$

Inserting this result into Eq. (10.78), we obtain

$$n = \frac{2m^3}{h^3}\frac{2\pi kT}{m} \int_{u=\sqrt{2\epsilon_{\text{surface}}/m}}^{\infty} u\,du \ln[1 + e^{(\epsilon_0 - mu^2/2)/kT}]. \tag{10.84}$$

Changing the variable of integration from u to $t = mu^2/2kT$, i.e. setting

$$u\,du = \frac{kT}{m}dt,$$

the integral becomes

$$n = \frac{4\pi mk^2 T^2}{h^3} \int_{\epsilon_{\text{surface}}/kT}^{\infty} dt \ln[1 + e^{-t}e^{\epsilon_0/kT}]. \tag{10.85}$$

In the next step we change the integration variable from t to $\tau = t - \epsilon_{\text{surface}}/kT$, so that

$$t = \tau + \frac{\epsilon_{\text{surface}}}{kT}.$$

The equation then becomes

$$n = \frac{4\pi mk^2 T^2}{h^3} \int_0^{\infty} d\tau \ln[1 + e^{-\tau}e^{(\epsilon_0 - \epsilon_{\text{surface}})/kT}]. \tag{10.86}$$

In the next step we use an integration by parts (with $\mu = e^{(\epsilon_0 - \epsilon_{\text{surface}})/kT}$):

$$\int d\tau \ln[1 + \mu e^{-\tau}] = \tau \ln[1 + \mu e^{-\tau}] + \int \frac{\tau d\tau}{1 + \mu e^{-\tau}} \mu e^{-\tau} = \tau \ln[1 + \mu e^{-\tau}] + \int \frac{\tau d\tau}{\mu^{-1} e^\tau + 1}. \quad (10.87)$$

Inserting this result into Eq. (10.86), we see that the first contribution vanishes at both limits, and the second contribution can be identified by Eq. (10.74) as a standard Fermi–Dirac integral, so that

$$n = \frac{4\pi m k^2 T^2}{h^3} F_1 \left(\frac{\epsilon_0 - \epsilon_{\text{surface}}}{kT} \right). \quad (10.88)$$

This result is examined further in Examples 10.8 and 10.9.

Example 10.8: Non-degenerate limit of thermionic emission

Show that in the non-degenerate (classical) limit of the result of Example 10.7 the classical result is obtained, *i.e.* Eq. (2.67),

$$n \simeq N \sqrt{\frac{kT}{2\pi m}} e^{\epsilon_{\text{surface}}/kT}. \quad (10.89)$$

Solution: In the classical limit $\eta = (\epsilon_0 - \epsilon_{\text{surface}})/kT \ll 1$, e^η small, and the unit term in the denominator of the Fermi–Dirac integral (10.74) can be neglected, *i.e.* (again using the factorial integral (12.151))

$$F_\kappa(\eta) = \int_0^\infty \frac{x^\kappa}{e^{x-\eta} + 1} dx \simeq \int_0^\infty x^\kappa e^{-x+\eta} dx = e^\eta \int_0^\infty x^\kappa e^{-x} dx = \kappa! e^\eta. \quad (10.90)$$

This reduces N of Eq. (10.77) where $\kappa = 1/2$ to (recall $(1/2)! = \sqrt{\pi}/2$ by Eq. (2.19))

$$N = \frac{8\sqrt{2\pi}(mkT)^{3/2}}{h^3} \left(\frac{1}{2} \right)! e^{\epsilon_0/kT} = \frac{4\sqrt{2}(\pi mkT)^{3/2}}{h^3} e^{\epsilon_0/kT}, \quad (10.91)$$

and n of Eq. (10.88) where $\kappa = 1$ to

$$n \simeq \frac{4\pi m k^2 T^2}{h^3} e^{(\epsilon_0 - \epsilon_{\text{surface}})/kT}, \quad (10.92)$$

so that

$$n \simeq N \sqrt{\frac{kT}{2\pi m}} e^{-\epsilon_{\text{surface}}/kT}, \quad (10.93)$$

which is the result obtained by classical considerations (*cf.* Eq. (2.67)).

Example 10.9: Thermionic emission under usual conditions in metals

Investigate the result of Example 10.7 under usual conditions in metals, *i.e.* for $\eta \gg 1$.

Solution: In metals $\epsilon_0/kT \gg 1$, and it is then necessary to use the approximation valid for $\eta \gg 1$, *i.e.*

$$\frac{1}{e^{x-\eta} + 1} \simeq \left\{ \begin{array}{ll} 1 & \text{for } x < \eta \\ 0 & \text{for } x > \eta \end{array} \right\}, \quad \text{giving } F_\kappa(\eta) \approx \int_0^\eta x^\kappa dx = \frac{\eta^{\kappa+1}}{\kappa + 1}. \quad (10.94)$$

Thus

$$N \overset{(10.77)}{\simeq} \frac{8\sqrt{2}\pi(mkT)^{3/2}}{h^3} \frac{(\epsilon_0/kT)^{3/2}}{3/2} = \frac{16\sqrt{2}\pi m^{3/2}\epsilon_0^{3/2}}{3h^3}, \tag{10.95}$$

which is the relation (10.10) determining the Fermi energy ϵ_0. On the other hand,

$$\frac{\epsilon_0 - \epsilon_{\text{surface}}}{kT} \ll 1,$$

and the approximation for $\eta \ll 1$ must be used. In this case as before in Eq. (10.92):

$$n \simeq \frac{4\pi mk^2 T^2}{h^3} e^{(\epsilon_0 - \epsilon_{\text{surface}})/kT}, \tag{10.96}$$

but now with a different value for ϵ_0. The quantity $\phi := \epsilon_{\text{surface}} - \epsilon_0$ is often called the "*work function*", being the work required to move an electron, which initially has the Fermi energy ϵ_0, across the potential barrier at the metallic surface.

10.4 Problems without Worked Solutions

Example 10.10: Relating Fermi–Dirac to Bose–Einstein integrals
Defining Fermi–Dirac and Bose–Einstein integrals respectively by the expressions

$$F_p(\eta) = \frac{1}{p!} \int_0^\infty \frac{\epsilon^p d\epsilon}{e^{\epsilon-\eta} + 1} \quad \text{and} \quad B_p(\eta) = \frac{1}{p!} \int_0^\infty \frac{\epsilon^p d\epsilon}{e^{\epsilon-\eta} - 1}, \tag{10.97}$$

show that

$$F_p(\eta) = B_p(\eta) - \frac{1}{2^p} B_p(2\eta). \tag{10.98}$$

Hence show that one can form the formal expansion[§§]

$$B_p(\eta) = \sum_{s=0}^\infty (2^{-p})^s F_p(2^s \eta). \tag{10.99}$$

Example 10.11: Mean energy of electrons
How does the energy of electrons in a metal vary with volume at constant density N/V, and how with N at constant volume? Can one obtain a sensible value of a mean energy of an electron as in the case of an ideal gas? What is the explanation? [Hint: Recall Pauli's principle].

Example 10.12: Temperature dependence of entropy
Without detailed calculations explain: (a) How does the entropy of electrons in a metal vary with temperature, and (b) how does the entropy of the radiation in a cavity vary at high temperatures?

Example 10.13: Fermi gas in a magnetic field
Consider a perfect Fermi gas consisting of spin-1/2 particles in a constant magnetic field B in a volume V. The energy of a particle is

$$\epsilon = \frac{p^2}{2m} - 2\mu_B B s_z, \quad s_z = \pm\frac{1}{2}. \tag{10.100}$$

Calculate at least to leading order the energy E of the gas and evaluate $N = \sum_i n_i$.

[§§]J. Clunie [13], R.B. Dingle [19], [20].

Example 10.14: Pressure of a Fermi gas

In Example 7.3 it was shown that in all 3 statistics $PV = 2E/3$ for a perfect gas of massive particles with energy $\epsilon = p^2/2m$. What is in the limit $T \to 0$ the value of E/N? Hence show that in the case of a nonrelativistic perfect Fermi gas the pressure P in the limit $T \to 0$ is

$$P = \frac{2}{5}\epsilon_0. \tag{10.101}$$

How does this differ from the case of a Bose–Einstein gas? What are the implications of this result for solids and fluids?

Example 10.15: Fermi momentum

Show that at temperature $T \simeq 0$ the momentum of the fastest electrons on top of the Fermi distribution is

$$p_F \simeq \frac{\hbar}{d}\pi, \quad v_F \simeq \frac{\hbar}{md}\pi, \tag{10.102}$$

where d is the lattice distance.[¶¶] Assuming $d = n$ Bohr radii, show that the corresponding velocity is $v_F = 7 \times 10^6/n$ meters/second.

Example 10.16: Evaluation of Fermi energy

Evaluate the Fermi energy of Eq. (10.10) for one ion in a cube of volume $(d/2)^3$ and show that

$$\epsilon_0 \simeq \frac{\hbar^2}{2md^2}. \tag{10.103}$$

Example 10.17: An electron pair at the Fermi surface

The wave function of the relative motion of a pair of free electrons of mass m and moving against each other with relative momentum p as for a Cooper pair on or very near the Fermi surface in a large sphere of radius R is given by

$$\psi(p) = \sqrt{\frac{2}{R}} \sin\left(\frac{pr}{\hbar}\right). \tag{10.104}$$

Show that the number of S-states in an energy interval $d\epsilon$ near the Fermi surface is

$$dn = \left(\frac{mR}{2\pi\hbar p_F}\right)d\epsilon. \tag{10.105}$$

Example 10.18: The electron motion in a superconductor

In superconductivity the motion of an electron on top of the Fermi distribution, *i.e.* at $T = 0$, attracts the positive ions of the lattice (here taken with charge $+1$ and lattice separation d) towards its path over a distance l and breadth δ behind it. Using the considerations of Examples 9.12, 10.15 and 10.16, show that

$$\delta \simeq \sqrt{\frac{m_e}{M}}d, \quad l \simeq \sqrt{\frac{M}{m_e}}d, \tag{10.106}$$

where M is the mass of the ion, and m_e that of the electron. [84] (Note the difference between this displacement of the ions and their vibration).
[Hint: Start from $p = M(\delta\omega_D)$, p = ion momentum $\simeq e^2/v_e d$, v_e = electron velocity, and $l = v_e/\omega_D$].

[¶¶]V.F. Weisskopf [84].

Example 10.19: Number of Cooper pairs

(a) In a superconducting metal Cooper pairs are formed as bound states of two electrons in a thin slice of thickness Δ in energy at the top of the Fermi distribution with energy E_F. If n and n' denote respectively the number of electrons and the number of Cooper pairs per unit volume, show that

$$n' \simeq \frac{3}{4} \frac{\Delta}{E_F} n. \tag{10.107}$$

[Hint: Recall that the number of states expressed in energy is proportional to $\sqrt{E} dE$].

(b) What is the distance between Cooper pairs if $n' = 1.25 \times 10^{-4}$ per cm^3, d being the nearest distance between neighbouring ions.

(Answer: $20d$).

Chapter 11

Limitations of the Preceding Theory — Improvement with Ensemble Method

11.1 Introductory Remarks

In this chapter we pinpoint weak aspects of the theory developed in the preceding chapters. We then introduce the concept of *ensemble* (or assembly) and show how this helps to improve the earlier arguments, although the results for the distribution functions — again derived by maximization — are the same as before.

The theory developed in the preceding chapters has three main faults:

1. The first fault is concerned with the physical conditions to which the system of interest is subjected. In previous work the energy

$$E = \sum_i n_i \epsilon_i$$

was assumed to be known exactly. This is true only if the system of interest is completely isolated. In practice no system is completely isolated. In actual practice the system of interest is normally surrounded by a large number of objects in thermal contact with it. This is depicted schematically in Fig. 11.1. There the temperature of the large heat bath is regarded to be constant because when the flask loses heat the temperature of the surrounding heat bath is not appreciably altered. Hence the *temperature* of the system of interest is constant, not the energy E! The above system is a *closed system* because the

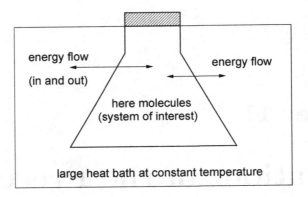

Fig. 11.1 A flask containing the system, the flask itself immersed in a heat bath.

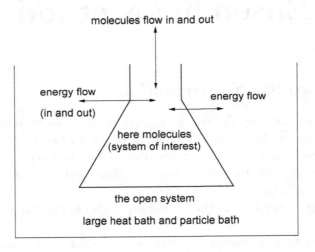

Fig. 11.2 The open system with heat and molecules flowing in and out.

number of conserved elements is constant. An *open* system is shown
schematically in Fig. 11.2.

2. Previously we used the method of maximum probability and found that
   particular set of occupation numbers n_i for which W, the number of
   arrangements of the elements of the system, was a maximum, *i.e.* we
   found that set which occurred most often as illustrated in Fig. 11.3.
   Consider one particular one-particle energy level ϵ_i only. Clearly we
   should have used the *mean occupation numbers* $\bar{n}_i$. These are pretty
   well the same as those at the maximum if the graph is sharp as in
   Fig. 11.4. But then we must prove that the peak is sharp.

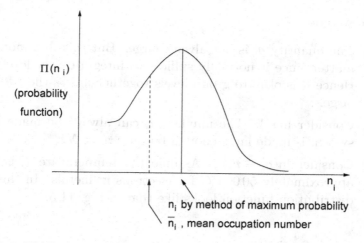

Fig. 11.3 The mean and maximum occupation numbers of energy level ϵ_i.

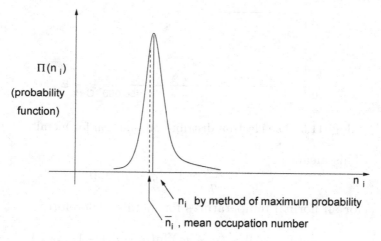

Fig. 11.4 The sharp maximum.

3. The third defect of the preceding considerations is that we used bad mathematics. We have used the *Stirling approximation*.

$$\ln n! \propto n(\ln n - 1). \tag{11.1}$$

This is reasonably okay if n is very large. Consider the case of Fermi–Dirac statistics:

$$W_{FD} = \prod_i \frac{g_i!}{(g_i - n_i)!n_i!}. \tag{11.2}$$

Here we note:

(a) The quantity g_i is not always large. But this does not normally matter since it normally suffices to integrate over levels anyway. Hence it is okay to group levels together unless the effective g_i is large.

(b) Consider $\ln n_i!$. The number n_i can always be made big if the system is made large enough because $n_i \propto N$.

(c) Consider $\ln(g_i - n_i)!$. Assume the temperature T is less than approximately $50\,000°C$ for electrons in metals. In this case the distribution function looks like that in Fig. 11.5.

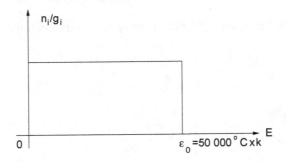

Fig. 11.5 The electron distribution function for metals.

This means:

$$\frac{n_i}{g_i} \simeq 1, \quad g_i - n_i \simeq 0 \tag{11.3}$$

for all normal temperatures for metals. Therefore

$$\ln(g_i - n_i)! \sim (g_i - n_i)[\ln(g_i - n_i) - 1] \quad \text{fails!} \tag{11.4}$$

In the following we show how these objections can be circumvented. The casual reader is warned to note the difference between ϵ_i in previous considerations and E_i from later on here (see Sec. 11.3).

11.2   Ensembles — Three Types

11.2.1   Ensembles and ergodic hypothesis

We introduce a large number of replicas of the system of interest as depicted for a closed system in Fig. 11.6, each of which is subjected to the same physical conditions as the system of interest. The set of all these *macroscopic*

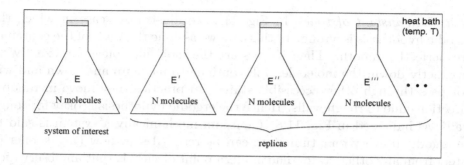

Fig. 11.6 Replicas of the closed system of interest with energies different to that
of the system immersed in a heat bath: The *canonical ensemble.*

systems is called an *ensemble.* The replicas can have at any instant of time
a slightly different energy to that of the system of interest. The *ensemble
distribution function* ρ, $\rho(E)$, is defined to be proportional to the number of
replica systems having a given property — in the present case the number
having the energy E in the given heat bath (thus $\rho(E')$ is the number of
replica systems in the same heat bath, however with energy E', so that ρ is a
property of the particular heat bath). If the system of interest is open, then
$\rho = \rho(E, N)$ since N could then vary as well.

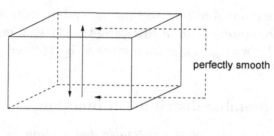

start molecules off going exactly up,
or exactly down

Fig. 11.7 Failure of the ergodic hypothesis for smooth walls.

We now make a hypothesis (due to Boltzmann), called the *ergodic hypothesis*
(*ergodic* from $\epsilon\rho\gamma o\nu$, meaning work or energy). This hypothesis says: In the
course of time the system of interest will pass through all states represented
by various replica systems. In other words, *the average over time* (the time
average of a dynamical variable is obtained by taking the average over phase
displacements* or periods) of the behaviour of the system of interest is the
same as the average over the ensemble of the behaviour of the replica systems

---

*See A. Messiah [43], Sec. 12.2.3.

at a given instant of time. In Fig. 11.7 we illustrate a case in which the ergodic hypothesis is wrong. In this case we assume the walls of the container are perfectly smooth. Then if we start the molecules off going exactly up or exactly down, the molecules will continue to move up and down and will not pass through other accessible states. In practice they move in random directions due to rough walls. Hence the ergodic hypothesis is okay for rough walls, as indicated in Fig. 11.8. Quantum mechanically a system is said to be ergodic if every quantum state can be reached somehow (*i.e.* sooner or later) from any other state. The modern point of view is that an observation

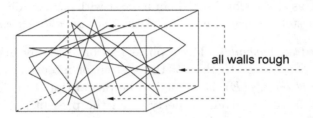

all walls rough

Only in an infinitely long period of time does the classical
system pass through every accessible position on the energy surface

Fig. 11.8 The ergodic hypothesis for rough walls.

is made in a finite interval of time so that the system cannot reach every accessible position or state — and in this sense the classical ergodic hypothesis is considered to be wrong.[†] The time average is replaced by the ensemble average.[‡]

11.2.2   The ensemble distribution function

We now proceed to determine the *ensemble distribution function ρ*. We consider three cases:

1. For an *isolated system* the energy E and the total number of particles N are both known (*i.e.* constant). Therefore every replica system in the ensemble must have the same values of total energy E and total number of particles N; the positions (classically points in phase space) and hence states of the individual particles may change. This isolated system (of isolated systems) is called the *microcanonical ensemble.* In this case the function ρ is the same for every replica system as

---

[†]This topic deserves more discussion than we can enter into here. Therefore see *e.g.* F. Reif [59], Sec. 15.14.

[‡]See *e.g.* M. Toda, R. Kubo and N. Saito [77], Chapter 5.

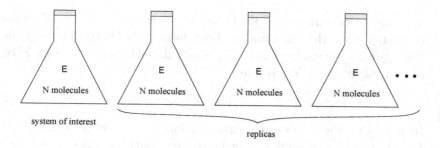

Fig. 11.9 Replicas of the isolated system of interest — each with
the same energy E: The *microcanonical ensemble*.

indicated in Fig. 11.9. Formulated quantum mechanically this says:
The isolated system in equilibrium occupies each of its accessible states
(classically phase space points, *cf.* Fig. 11.10) with the same probability (*cf.* the same probabilities for each face of a die). This statement
is also described as the *fundamental postulate of equal a priori probabilities* of an isolated system. Recalling our earlier definition of the *a
priori* probability of Chapter 3 and Secs. 4.2 and 6.2, we see now that
this fundamental postulate may be considered as equating the *a priori* probability to the degeneracy of a system (the number of different
states with the same energy). Observe that no temperature is involved.
This isolated system with fixed energy and no influence from outside
is said to be in a *pure state*, and since this is only one state that it
occupies its entropy (logarithm of number of accessible states) is zero.

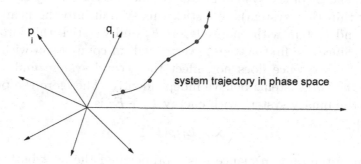

Fig. 11.10 The classical ergodic equivalent to the ensemble of Fig. 11.9:
The time-trajectory in phase space.

2. Next we consider a *closed system in a large heat bath*. In this case
   the number of particles N is known. Therefore every replica system

has this same number of particles N. However, in view of exchange of energy with the surrounding heat bath (*i.e.* thermal energy, heat), the replica systems have energies which differ from the energy E of the system of interest. Therefore

$$\rho = \rho(E). \tag{11.5}$$

In principle the replica systems can have any energy as indicated in Fig. 11.6. This ensemble in a large heat bath is called the *canonical ensemble*. Thus quantum mechanical stationary states are not exact and become socalled *mixed states* in view of thermal interaction with the surrounding heat bath.

3. Finally we consider an *open system in a large heat and element bath.* Here neither N nor E is known. Therefore

$$\rho = \rho(E, N). \tag{11.6}$$

This ensemble is called the *grand canonical ensemble.*

The most important of the three cases is case 2 which we therefore consider in detail.[§]

11.3   The Canonical Ensemble of a Closed System

In the case of the canonical ensemble the problem is to find $\rho(E)$ for the system in a large heat bath. We take one system with energy levels E_i (this is the total energy of the system in state i, and not the energy of a particular element within that system). We stick this system into the heat bath and take a second system with energy levels $\tilde{E}_j$ and we stick this into the heat bath also. Since the first system is very small in comparison with the total heat bath, its presence does not affect the second system and *vice versa*. Therefore the two systems behave independently. Therefore the probability of finding the binary system with energy $E_i + \tilde{E}_j$ is

$$\propto \rho(E_i)\rho(\tilde{E}_j).$$

Here we have the same ρ's since ρ is a property of the heat bath, and both systems are in the same heat bath. Consider this binary system as a new system within the heat bath. The probability of finding this system of energy

$$E = E_i + \tilde{E}_j$$

[§]One can actually establish the equivalence of the three cases but we do not enter into this here.

is proportional to

$$\rho(E) = \rho(E_i + \tilde{E}_j).$$

Since $\rho = \rho(\text{total energy})$, this must be independent of the individual E_i and $\tilde{E}_j$, and depend only upon their sum $E = E_i + \tilde{E}_j$ independently of the particular values of E_i and $\tilde{E}_j$. Therefore

$$\rho(E_i + \tilde{E}_j) \propto \rho(E_i)\rho(\tilde{E}_j) \qquad (11.7)$$

independently of the particular values of E_i and $\tilde{E}_j$. This can only be satisfied for a function of the type

$$\rho(E) \propto (\text{something})^E.$$

For mathematical convenience we write

$$\text{something} = e^{-\mu},$$

where μ is as yet unknown. Therefore

$$\rho(E) \propto e^{-\mu E}. \qquad (11.8)$$

We check this: The quantities

$$\rho(E_i + \tilde{E}_j) \propto e^{-\mu(E_i + \tilde{E}_j)}, \quad \rho(E_i)\rho(\tilde{E}_j) \propto e^{-\mu E_i} e^{-\mu \tilde{E}_j}$$

are seen to be equal.

11.3.1 Thermodynamics of a closed system in a heat bath

We normalize the *ensemble distribution function* $\rho(E)$ (different replica systems of the given system having different energies E, *i.e.* $E', E'', \ldots$, with energy levels $E_i', E_i'', \ldots$) to make it into a real probability, *i.e.* so that $\sum_i \rho(E_i) = 1$, where the sum is taken over all possible total energy levels of the system of interest (eigenvalues). Therefore we have normalized

$$\rho(E_i) = \frac{e^{-\mu E_i}}{\sum_j e^{-\mu \tilde{E}_j}}. \qquad (11.9)$$

We now consider the quantity

$$S = -k \ln \rho, \qquad (11.10)$$

and we show that the average $\overline{S}$ corresponds to the thermodynamical concept of *entropy*,

$$\overline{S} = -k\overline{\ln \rho}, \qquad (11.11)$$

since the average over the ensemble replaces by the ergodic hypothesis the average over time of the behaviour of the system of interest. We have:

$$\overline{S} = -k\overline{\ln\rho} = -k\sum_i \rho(E_i)\ln\rho(E_i)$$

$$= -k\sum_i \rho(E_i)\left(\ln\frac{e^{-\mu E_i}}{\sum_j e^{-\mu E_j}}\right)$$

$$= \mu k\sum_i \rho(E_i)E_i + k\left(\ln\sum_i e^{-\mu E_i}\right)\sum_j \rho(E_j), \qquad (11.12)$$

where $\sum_i \rho(E_i)E_i$ is the average or mean energy $\overline{E}$ in the present context defined by[1]

$$\overline{E} = \frac{\sum_i \rho(E_i)E_i}{\sum_i \rho(E_i)} \quad\text{with}\quad \sum_i \rho(E_i) = 1. \qquad (11.13)$$

Thus we have

$$\overline{S} = \mu k\overline{E} + k\ln\sum_i e^{-\mu E_i}. \qquad (11.14)$$

We compare this equation with the thermodynamic equation for the free energy $\overline{F}$ (see Eq. (5.10)):

$$F = \overline{E} - T\overline{S} \quad\text{(all average values),}\quad \text{or}\quad \overline{S} = \frac{\overline{E}}{T} - \frac{\overline{F}}{T}. \qquad (11.15)$$

Thus, by comparison,

$$\mu k = \frac{1}{T}, \quad \overline{F} = -kT\ln\sum_i e^{-\mu E_i}, \qquad (11.16)$$

and

$$\mu = \frac{1}{kT}. \qquad (11.17)$$

Alternatively we consider the second law of thermodynamics, Eq. (5.15),

$$Td\overline{S} = d\overline{E} + \overline{P}dV, \qquad (11.18)$$

which implies, that from this equation the *temperature* T is defined by

$$T = \left(\frac{\partial\overline{E}}{\partial\overline{S}}\right)_V = \frac{1}{(\partial\overline{S}/\partial\overline{E})_V}. \qquad (11.19)$$

---

[1]This is a different form from our earlier relation (5.23), *i.e.* $\overline{E} = -\partial\ln Z/\partial\mu$ for $Z = \sum_i e^{-\mu E_i}$.

The energy $\overline{E}$ depends only on the volume. Therefore during the differentiation the expression

$$k \ln \sum_i e^{-\mu E_i}$$

is constant. Therefore from Eq. (11.14)

$$\left(\frac{\partial \overline{S}}{\partial \overline{E}}\right) = \mu k, \tag{11.20}$$

and therefore[||] μ in Eq. (11.9) can be replaced by $1/kT$ from Eq. (11.17),

$$\rho(E) = \frac{e^{-E_i/kT}}{\sum_i e^{-E_i/kT}}, \tag{11.21}$$

and also in $\overline{F}$,

$$\overline{F} = \overline{E} - T\overline{S} = -kT \ln \sum_i e^{-E_i/kT}. \tag{11.22}$$

Or we can call this the *mean free energy* — this is a matter of definition. The entropy $\overline{S}$ here in this classical consideration (identified with that in the second law of thermodynamics) is also called *coarse grained entropy* in order to distinguish it from the quantum mechanical entropy of Sec. 11.6.

11.4   The Grand Canonical Ensemble

One can prove (see Examples 11.4, 11.5) that correspondingly we find for an *open system* that

$$\rho(E_i, N) = \frac{\exp[(N\epsilon_0 - E_i)/kT]}{\sum_i \sum_j \exp[(N_j\epsilon_0 - E_i)/kT]}, \tag{11.23}$$

where the two sums extend over all possible energies E_i and all possible numbers N_j, ϵ_0 being the *chemical potential*. We note that

$$e^{-E/kT}$$

is the *energy weighting factor* for each arrangement when the system of interest is in a large heat bath, and

$$e^{N\epsilon_0/kT}$$

is the *population weighting factor* when the system of interest is open.

[||] *Cf.* A. Messiah [43], Vol. I, Sec. 12.2.5.

11.5   Ensemble Method of Maximum Probability

In the present ensemble consideration we must attach a factor proportional
to

$$e^{-E/kT}$$

to each possible *arrangement* of the elements in the system. This factor is
called the *energy weighting factor*. We had before that (*cf.* Eq. (6.44))

$$W_{FD} = \prod_i \frac{g_i!}{n_i!(g_i - n_i)!}. \tag{11.24}$$

Attach the exponential factor to each of these arrangements. But note that

$$e^{-E/kT} = e^{-\sum_i n_i \epsilon_i / kT} = \prod_i e^{-n_i \epsilon_i / kT}. \tag{11.25}$$

Therefore the *weighted number of arrangements* is:

$$\tilde{W}_{FD} \propto \prod_i \frac{g_i! e^{-n_i \epsilon_i / kT}}{n_i!(g_i - n_i)!}. \tag{11.26}$$

We now proceed as for the isolated system.
1. *Maximization of $\tilde{W}_{FD}$*:
    Instead of considering the derivative of $\tilde{W}_{FD}$, it is easier to consider that
of the logarithm, *i.e.*

$$d \ln \tilde{W}_{FD} = 0, \quad \tilde{W}_{FD} \propto \prod_i \frac{g_i! e^{-n_i \epsilon_i / kT}}{n_i!(g_i - n_i)!}. \tag{11.27}$$

Therefore:

$$\ln \tilde{W}_{FD} \propto \sum_i [\ln g_i! - n_i \epsilon_i / kT - \ln n_i! - \ln(g_i - n_i)!]$$

$$= \sum_i [\ln g_i! - n_i \epsilon_i / kT - n_i(\ln n_i - 1)$$

$$- (g_i - n_i)\{\ln(g_i - n_i) - 1\}]. \tag{11.28}$$

Therefore the derivative is

$$d \ln \tilde{W}_{FD} \propto \sum_i dn_i [-\epsilon_i / kT - (\ln n_i - 1) - 1 + 1$$

$$+ \{\ln(g_i - n_i) - 1\}]$$

$$= -\sum_i [\epsilon_i / kT + \ln n_i - \ln(g_i - n_i)]dn_i$$

$$= -\sum_i \ln \frac{n_i e^{\epsilon_i / kT}}{g_i - n_i} dn_i. \tag{11.29}$$

2. *Subsidiary condition*:

Here in the ensemble consideration we have only one subsidiary condition, whereas in Chapter 7 we had two. We have

$$\sum_i n_i = N \quad \text{for conserved elements.} \tag{11.30}$$

(We have no subsidiary condition for E). Therefore

$$\sum_i dn_i = 0. \tag{11.31}$$

Hence by the method of *Lagrangian multipliers* (*cf.* Chapter 2, Eqs. (2.21), (2.22), (2.23)),

$$\ln \frac{n_i e^{\epsilon_i/kT}}{g_i - n_i} = -A', \quad A' = \text{const.}, \tag{11.32}$$

or

$$\frac{n_i e^{\epsilon_i/kT}}{g_i - n_i} = e^{-A'}, \qquad n_i e^{\epsilon_i/kT + A'} = g_i - n_i,$$

and hence**

$$n_i = \frac{g_i}{e^{\epsilon_i/kT + A'} + 1} \equiv \frac{g_i}{-A e^{\epsilon_i/kT} + 1}. \tag{11.33}$$

Thus one obtains the same distribution function as before in Chapter 7, now, however, on a basis avoiding the faults enumerated in Sec. 11.1. Fluctuations are actually different (the effect of the walls of the container have to be taken into account).

11.6   Comments on the Function ρ

The typical system we have in mind here is the ideal gas with energy that of noninteracting particles of mass m, *i.e.* particles with energy $\epsilon_i = p_i^2/2m$ in classical mechanics. Thus in such a case $E = \sum_i \epsilon_i = \sum_i p_i^2/2m$, and correspondingly $\rho(E) = \rho(p_1, p_2, p_3, \ldots, p_i, \ldots)$. Clearly in a more general case we have $\rho(E) = \rho(H(q_1, q_2, q_3, \ldots, p_1, p_2, p_3, \ldots)) = \rho(q_1, q_2, q_3 \ldots, p_1, p_2, p_3, \ldots)$ and still more generally ρ would be $\rho(q_1, q_2, q_3, \ldots, p_1, p_2, p_3, \ldots, t)$ with an explicit dependence on time t. Thus the function ρ depends on the time-dependent dynamical variables q_i, p_i (correspondingly in quantum mechanics

---

** Here in the method of maximum probability the parameter A' is a Lagrangian multiplier; in the method of steepest descents of Chapter 12 this parameter defines the position of maximum contribution to the contour integral. See the remarks in connection with Eq. (12.46).

on "states"), and therefore describes a motion — that of the ensemble. With no explicit t-dependence we have

$$\frac{\partial \rho}{\partial t} = 0, \quad \rho = \rho(E) \equiv \rho(H(q, p)). \tag{11.34}$$

This is the condition of *equilibrium*, the state of the system when disturbances have subsided. The *equation of motion of the ensemble*, called *Liouville equation*, is obtained by differentiation of ρ, *i.e.*:

$$\frac{d}{dt}\rho(q_1, q_2, q_3, \ldots, p_1, p_2, p_3, \ldots, t) = \sum_i \left(\frac{\partial \rho}{\partial q_i}\frac{dq_i}{dt} + \frac{\partial \rho}{\partial p_i}\frac{dp_i}{dt} \right) + \frac{\partial \rho}{\partial t}$$

$$= \sum_i \left(\frac{\partial \rho}{\partial q_i}\frac{\partial H}{\partial p_i} - \frac{\partial \rho}{\partial p_i}\frac{\partial H}{\partial q_i} \right) + \frac{\partial \rho}{\partial t}$$

$$= \{\rho, H\} + \frac{\partial \rho}{\partial t}, \tag{11.35}$$

where we used Hamilton's equations and $\{\cdots, \cdots\}$ is a Poisson bracket. If no additional replica systems are created, nor any annihilated, their number per unit volume ρ is constant, *i.e.*

$$\frac{d\rho}{dt} = 0. \tag{11.36}$$

The resulting equation is:

$$0 = \frac{\partial \rho}{\partial t} + \sum_i \left(\frac{\partial \rho}{\partial q_i}\dot{q}_i + \frac{\partial \rho}{\partial p_i}\dot{p}_i \right)$$

$$= \frac{\partial \rho}{\partial t} + \sum_i \left(\frac{\partial}{\partial q_i}(\rho\dot{q}_i) + \frac{\partial}{\partial p_i}(\rho\dot{p}_i) \right), \tag{11.37}$$

where

$$\rho\frac{\partial \dot{q}_i}{\partial q_i} + \rho\frac{\partial \dot{p}_i}{\partial p_i} = \rho\left[\frac{\partial}{\partial q_i}\left(\frac{\partial H}{\partial p_i}\right) + \frac{\partial}{\partial p_i}\left(-\frac{\partial H}{\partial q_i}\right) \right] = 0. \tag{11.38a}$$

Thus

$$0 = \frac{\partial \rho}{\partial t} + \boldsymbol{\nabla} \cdot (\rho\dot{\mathbf{v}}), \qquad \dot{\mathbf{v}} = (\dot{q}, \dot{p}). \tag{11.38b}$$

This is the *continuity equation* well known from other areas of physics (*e.g.* in electrodynamics for stationary currents). In the present context the equation is known as the *Liouville equation*;[††] it is the equation of motion of the

---

[††]Various forms of discussion can be found in the literature, for instance in F. Reif [59], Appendix A.13, H.J.W. Müller–Kirsten [48], Chapter 6, and [47], Chapter 2. For the context of electrodynamics see [50], Chapter 5.

ensemble of replica systems of a given system. The expression $\rho(q, p, t)\triangle q\triangle p$ is the number of systems of the ensemble at time t in the phase space volume element $\triangle q\triangle p$ with phase space coordinates in the intervals $q, q + \triangle q$ and $p, p + \triangle p$. Thus if $\partial\rho/\partial t = 0$, one has a stationary flow of systems, hence the term *stationary current* or *equilibrium*. Quantum mechanically the function ρ becomes an operator called the *density matrix*, and the above Poisson bracket becomes a commutator.

A system consisting of two subsystems (or more) which interact is said to be *entangled* or *correlated* (in quantum mechanics). The system which does not interact with any other system is said to be in a pure state, and its entropy is zero (one state). A system which interacts with another system is said to be in a mixed state; its entropy is nonzero. The eigenvalues of the density matrix have values between zero and one (the latter for a pure state). The entropy

$$\overline{S} = -k\mathrm{Tr}\overline{\rho}\ln\overline{\rho} \tag{11.39}$$

is then called *fine grained* or von Neumann entropy or *entanglement entropy*.

11.7   Applications and Examples

Example 11.1: Mean values of thermodynamical quantities

Defining the entropy of a system in a canonical ensemble as $S = -k\ln\rho(E)$, where $\rho(E)$ is the normalized ensemble distribution function, verify that[‡‡]

$$\left(\frac{\partial\overline{S}}{\partial\overline{E}}\right)_V = \frac{1}{T}, \quad \overline{F} = -kT\ln\left\{\sum_i e^{-E_i/kT}\right\},$$

$$\overline{P} = \left(T\frac{\partial\overline{S}}{\partial V}\right)_{\overline{E}} = -\left(\frac{\partial\overline{F}}{\partial V}\right)_T, \quad \text{and} \quad \overline{S} = -\left(\frac{\partial\overline{F}}{\partial T}\right)_V. \tag{11.40}$$

Solution: We start from the equations

$$S = -k\ln\rho(E), \quad \rho(E_i) = \frac{e^{-\mu E_i}}{\sum_j e^{-\mu E_j}}, \quad \mu = \frac{1}{kT}, \quad \overline{E} = \sum_i \rho(E_i)E_i. \tag{11.41}$$

[‡‡]Here we consider ensemble averages with $\overline{S}$ and $\overline{E}$ as below in the first line of Eq. (11.42) and in Eq. (11.41) respectively. The free energy F was defined in Eq. (5.10) as $\overline{E} - T\overline{S}$, so that there is really no need to write $\overline{F}$.

We begin with the average of the entropy (observe the difference between S and $\overline{S}$):

$$
\begin{aligned}
\overline{S} &= -k\overline{\ln \rho(E)} = -k\sum_i \rho(E_i)\ln \rho(E_i) = -k\sum_i \rho(E_i)\ln\left(\frac{e^{-\mu E_i}}{\sum_j e^{-\mu E_j}}\right)\\
&= -k\sum_i \rho(E_i)\left\{-\mu E_i - \ln\sum_j e^{-\mu E_j}\right\}\\
&= k\mu\sum_i \rho(E_i)E_i + k\sum_i \rho(E_i)\ln\sum_j e^{-\mu E_j}\\
&= \frac{1}{T}\overline{E} + k\ln\sum_j e^{-\mu E_j},
\end{aligned}
\tag{11.42}
$$

where we used $\sum_i \rho(E_i) = 1$. It follows that

$$
\left(\frac{\partial \overline{S}}{\partial \overline{E}}\right)_V = \frac{1}{T}.
\tag{11.43}
$$

But we have in thermodynamics the relation $F = E - TS$ which is really the averaged relation

$$
\overline{F} = \overline{E} - T\overline{S}, \qquad \text{or} \qquad \overline{S} = \frac{\overline{E}}{T} - \frac{\overline{F}}{T},
\tag{11.44}
$$

Thus by comparison with Eq. (11.42) we obtain

$$
\overline{F} = -kT\ln\left\{\sum_i e^{-E_i/kT}\right\}.
\tag{11.45}
$$

Next we differentiate this expression in order to evaluate the quantity:

$$
\begin{aligned}
-\left(\frac{\partial \overline{F}}{\partial T}\right)_V &= k\ln\sum_i e^{-\mu E_i} + kT\frac{\sum_i E_i(1/kT^2)e^{-\mu E_i}}{\sum_j e^{-\mu E_j}}\\
&= k\ln\sum_i e^{-\mu E_i} + \frac{1}{T}\frac{\sum_i E_i e^{-\mu E_i}}{\sum_j e^{-\mu E_j}}\\
&= k\ln\sum_i e^{-\mu E_i} + \frac{1}{T}\sum_i E_i \rho(E_i) \overset{(11.42)}{=} \overline{S}.
\end{aligned}
\tag{11.46}
$$

From the *macroscopic* thermodynamical relation $TdS = dE + PdV$, which really is the averaged relation

$$
Td\overline{S} = d\overline{E} + \overline{P}dV,
\tag{11.47}
$$

we obtain

$$
\overline{P} = T\left(\frac{\partial \overline{S}}{\partial V}\right)_{\overline{E}}.
\tag{11.48}
$$

From Eqs. (11.44) and (11.47) we obtain

$$
d\overline{F} = d\overline{E} - Td\overline{S} - \overline{S}dT = -\overline{P}dV - \overline{S}dT,
\tag{11.49}
$$

so that

$$
\left(\frac{\partial \overline{F}}{\partial V}\right)_T = -\overline{P}.
\tag{11.50}
$$

Example 11.2: Thermodynamical functions in terms of partition function

Express all the thermodynamical functions in terms of the *partition function* for a canonical ensemble, $Z = \sum_i e^{-E_i/kT}, 1/kT \equiv \mu$.

Solution: We have from Eq. (11.40):

$$\overline{F} = -kT \ln Z. \tag{11.51}$$

We also have $\rho(E_i) = e^{-\mu E_i}/Z$. Then

$$\overline{E} = \sum_i \rho(E_i)E_i = \frac{1}{Z}\sum_i E_i e^{-\mu E_i} = -\frac{1}{Z}\frac{\partial Z}{\partial \mu}. \tag{11.52}$$

It follows that

$$
\begin{aligned}
\overline{S} \;\overset{(11.42)}{=}\; & \frac{1}{T}\overline{E} + k\ln Z = -\frac{1}{TZ}\frac{\partial Z}{\partial \mu} + k\ln Z = -\frac{k}{TZ}\frac{\partial Z}{\partial(1/T)} + k\ln Z \\
= & -\frac{k}{TZ}(-T^2)\frac{\partial Z}{\partial T} + k\ln Z = \frac{kT}{Z}\frac{\partial Z}{\partial T} + k\ln Z = kT\frac{\partial}{\partial T}(\ln Z) + k\ln Z \\
= & \frac{\partial}{\partial T}(kT\ln Z),
\end{aligned}
\tag{11.53}
$$

which is in agreement with

$$\overline{S} = -\left(\frac{\partial \overline{F}}{\partial T}\right)_V = \frac{\partial}{\partial T}(kT\ln Z). \tag{11.54}$$

Example 11.3: Maxwell–Boltzmann distribution

Find by the method of maximum probability the distribution function for a Maxwell–Boltzmann system placed in a heat bath.

Solution: We consider first the application of the canonical ensemble method to Maxwell–Boltzmann statistics using the method of maximum probability. We must attach a factor proportional to $e^{-E/kT}$ to each possible arrangement of the elements in the system. In Maxwell–Boltzmann statistics the number of arrangements is given by Eq. (4.2),

$$W_{MB} = N! \prod_i \frac{g_i^{n_i}}{n_i!}. \tag{11.55}$$

Hence the *weighted number of arrangements* is

$$\tilde{W}_{MB} \propto N! \prod_i \frac{g_i^{n_i}}{n_i!}e^{-E/kT}, \quad E = \sum_j n_j \epsilon_j, \quad e^{-\sum_i n_i \epsilon_i/kT} = \prod_i e^{-n_i \epsilon_i/kT}. \tag{11.56}$$

Taking the logarithm we have

$$\ln \tilde{W}_{MB} \propto \ln \left\{ N! \prod_i \frac{g_i^{n_i}e^{-n_i \epsilon_i/kT}}{n_i!} \right\} = \sum_i \left[\ln N! + n_i \ln g_i - n_i \epsilon_i/kT - n_i(\ln n_i - 1) \right]. \tag{11.57}$$

Now maximize by forming the differential and equating this to zero:

$$
\begin{aligned}
d\ln \tilde{W}_{MB} &\propto \sum_i dn_i\{\ln g_i - \epsilon_i/kT - \ln n_i + 1 - 1\} = \sum_i dn_i\{\ln g_i - \ln n_i - \epsilon_i/kT\} \\
&= -\sum_i dn_i \left[\ln\left(\frac{n_i e^{\epsilon_i/kT}}{g_i}\right) \right] = 0.
\end{aligned}
\tag{11.58}
$$

We have one subsidiary condition, namely that for conserved elements,

$$N = \sum_i n_i, \qquad dN = \sum_i dn_i = 0. \tag{11.59}$$

Hence by the method of Lagrangian multipliers of Chapter 2 applied to the last two equations, we get

$$-\ln\left(\frac{n_i e^{\epsilon_i/kT}}{g_i}\right) = \text{const.} \equiv A'. \tag{11.60}$$

For nonconserved elements this constant vanishes. Hence:

$$n_i e^{\epsilon_i/kT} = g_i e^{-A'}, \qquad n_i = \frac{g_i}{e^{A'+\epsilon_i/kT}}, \tag{11.61}$$

and hence

$$n_i = \frac{g_i}{e^{(\epsilon_i - \epsilon_0)/kT}}, \tag{11.62}$$

where $-\epsilon_0/kT = A'$. We recognize the result as that obtained earlier in Eq. (4.13). The case of Fermi–Dirac statistics has been dealt with above in the text. The case of Bose–Einstein statistics proceeds correspondingly.

Example 11.4: Two open systems in heat and element bath

By considering two open systems placed in the same large heat and element bath, one with energy E and population N, and the other with energy E' and population N', show that

$$\rho(E + E', N + N') \propto \rho(E, N)\rho(E', N'), \tag{11.63}$$

assuming that the ensemble distribution function is unique for a given large heat and element bath. Hence deduce that the distribution function for a grand canonical ensemble is

$$\rho(E, N) = \frac{e^{\alpha N - \mu E}}{\sum_i \sum_j e^{\alpha N_j - \mu E_i}}, \tag{11.64}$$

where $\sum_i \sum_j$ refers to summation over both all attainable energy eigenvalues E_i and all possible populations N_j.

Solution: We take one system with total energy E and population N and another system with energy E' and population N'. Since each of the systems is very small in comparison with the total heat bath, either system does not affect the other. Hence the two systems behave independently. Therefore the probability of finding the binary system with energy $E + E'$ and population $N + N'$ is proportional to

$$\rho(E, N)\rho(E', N'),$$

assuming that the ensemble distribution function is unique. Here we have the same ρ's since ρ is a property of the heat bath and both systems are in the same heat bath.

We now consider this binary system as a new system within the heat bath. The probability of finding this system of energy $E^* = E + E'$ and population $N^* = N + N'$ is proportional to

$$\rho(E^*, N^*) = \rho(E + E', N + N').$$

This is independent of individual E, E', N, N', and depends only on the sums $E + E', N + N'$. It follows that

$$\rho(E + E', N + N') \propto \rho(E, N)\rho(E', N'). \tag{11.65}$$

This relation can only be satisfied for a function of the type

$$\rho(E, N) \propto (\text{something})^N (\text{something else})^E.$$

We put

$$\text{something} = e^{\alpha}, \quad \text{something else} = e^{-\mu},$$

for mathematical convenience. Next we normalize the function ρ to convert it into a proper probability, *i.e.* we set $\sum\sum \rho = 1$, where the sums are taken over all possible N and E. Then

$$\rho(E, N) = \frac{e^{\alpha N - \mu E}}{\sum_i \sum_j e^{\alpha N_j - \mu E_i}}. \tag{11.66}$$

Example 11.5: Thermodynamics of a grand canonical ensemble

Defining the entropy of a system in a grand canonical ensemble as $S = -k \ln \rho$, where $\rho(E, N)$ is the normalized ensemble distribution function, determine the parameters α and μ of Example 11.4 by evaluating $(\partial \overline{S}/\partial \overline{E})_{\overline{N}, V}$ and $(\partial \overline{S}/\partial \overline{N})_{\overline{E}, V}$, and comparing with the corresponding values obtained from the thermodynamical relation

$$T d\overline{S} = d\overline{E} + \overline{P} dV - \epsilon_0 d\overline{N} \tag{11.67}$$

valid for an open system, ϵ_0 being the *chemical potential* (thermodynamic potential per element).[§§]

Solution: Assuming the definition $S = -k \ln \rho$, we have

$$\overline{S} = -k \overline{\ln \rho} = -k \sum \sum \rho \ln \rho = -k \sum \sum \rho \ln \frac{e^{\alpha N - \mu E}}{\sum_j \sum_i e^{\alpha N_j - \mu E_i}}$$

$$= -k \sum \sum \rho \left[\alpha N - \mu E - \ln \sum_j \sum_i e^{\alpha N_j - \mu E_i} \right]$$

$$= -k\alpha \sum_j \rho N_j + \mu k \sum_i \rho E_i + k \ln \left[\sum_j \sum_i e^{\alpha N_j - \mu E_i} \right] \sum \sum \rho. \tag{11.68}$$

But

$$\sum \sum \rho = 1, \quad \sum \sum \rho E = \overline{E}, \quad \sum \sum \rho N = \overline{N}. \tag{11.69}$$

Therefore

$$\overline{S} = -k\alpha \overline{N} + \mu k \overline{E} + k \ln \sum_j \sum_i e^{\alpha N_j - \mu E_i}. \tag{11.70}$$

From this equation we deduce that

$$\left(\frac{\partial \overline{S}}{\partial \overline{E}} \right)_{\overline{N}, V} = \mu k, \tag{11.71a}$$

and

$$\left(\frac{\partial \overline{S}}{\partial \overline{N}} \right)_{\overline{E}, V} = -k\alpha. \tag{11.71b}$$

Also:

$$T d\overline{S} = d\overline{E} + \overline{P} dV - \epsilon_0 d\overline{N}, \tag{11.72}$$

Therefore

$$d\overline{S} = \frac{d\overline{E}}{T} + \frac{\overline{P} dV}{T} - \frac{\epsilon_0 d\overline{N}}{T}. \tag{11.73}$$

[§§]The extended quantity $\epsilon_0 + q\phi$, where q is the charge of a particle and ϕ an electrostatic potential, is called *electrochemical potential* — see *e.g.* P.T. Landsberg [39], pp.320-324.

Therefore:

$$\left(\frac{\partial \overline{S}}{\partial \overline{E}}\right)_{\overline{N},V} = \frac{1}{T}, \tag{11.74a}$$

and

$$\left(\frac{\partial \overline{S}}{\partial \overline{N}}\right)_{\overline{E},V} = -\frac{\epsilon_0}{T}. \tag{11.74b}$$

Hence from Eqs. (11.71a), (11.74a), (11.71b), (11.74b) we obtain:

$$\frac{1}{T} = \mu k, \quad \mu = \frac{1}{kT}, \quad \text{and} \quad -\frac{\epsilon_0}{T} = -k\alpha, \quad \alpha = \frac{\epsilon_0}{kT}. \tag{11.75}$$

Thus finally

$$\rho(E,N) = \frac{e^{(N\epsilon_0 - E)/kT}}{\sum_j \sum_i e^{(N_j\epsilon_0 - E_i)/kT}} \quad \text{and} \quad \overline{P} = T\left(\frac{\partial \overline{S}}{\partial V}\right)_{\overline{E},\overline{N}}. \tag{11.76}$$

11.8   Problems without Worked Solutions

Example 11.6: Bose–Einstein distribution
Find by the method of maximum probability (*i.e.* parallel to the method of Example 11.3) the distribution function for a Bose–Einstein system placed in a heat bath.

Example 11.7: The ergodic hypothesis and the linear harmonic oscillator
The classical ergodic hypothesis claims that the phase space trajectory (traced by $(q(t), p(t))$) of a system passes through each point of the hyperplane $E = H(q,p)$. Show that the classical linear harmonic oscillator with

$$H(q,p) = \frac{p^2}{2m} + \frac{1}{2}m\omega^2 q^2 \tag{11.77}$$

satisfies this claim exactly.
[Hint: Show that $q = \sqrt{2E/m\omega^2}\,\sin(\omega t + \delta), p = \sqrt{2mE}\,\cos(\omega t + \delta)$, and that after period $2\pi/\omega$ every point of the hyperplane has been passed through].

Example 11.8: Mean pressure in extreme relativistic case
Consider a perfect quantum gas in a volume V and determine the relation between its mean energy $\overline{E}$ and the mean pressure $\overline{P}$ in both the nonrelativistic and extreme relativistic cases using $\overline{P} = kT\partial \ln Z/\partial V, Z = \sum_i \exp(-E_i/kT)$ of Eq. (11.40). (Answer: $\overline{P}V = (2/3, 1/3)\overline{E}$).

Example 11.9: Difference between entropies
Boxes A and B are separated by a common partition with a tiny hole. With an explosive device A is filled with photons which then gradually escape through the hole into B. Sketch roughly the behaviour of the difference of coarse grained and fine grained entropies of B as a function of the fraction of photons escaped from A.
[Hint: B starts in a pure state and also ends in a pure state, see [73], p.76].

Example 11.10: Entropy as measure of entanglement
The von Neumann entropy S is given by the relation

$$S = -k\mathrm{Tr}\rho\ln\rho = -k\Sigma_i\rho_i\ln\rho_i, \tag{11.78}$$

and provides a measure of the departure from a pure state (in which case $S = 0$). An eigenvalue (*i.e.* diagonalized element) ρ_i describes the probability that the system represented by the density matrix is in state i. What is (a) the von Neumann entropy in the case of the diagonalized density matrix

$$\rho = \frac{1}{2} \begin{pmatrix} 1 & 0 \\ 0 & 1 \end{pmatrix} \qquad (11.79)$$

and (b) the number of states in the ensemble? (Answers: (a) $k \ln 2 = 0.69$, (b) 2).

Example 11.11: Density matrix of a pure state

Show by diagonalization[¶¶] that the matrix

$$\rho = \frac{1}{2} \begin{pmatrix} 1 & 1 \\ 1 & 1 \end{pmatrix} \qquad (11.80)$$

represents a pure state (*i.e.* has eigenvalues $1, 0$).

[¶¶] For a revision of diagonalization see *e.g.* [48], p.273.

Chapter 12

Averaging instead of Maximization, and Bose–Einstein Condensation

12.1 Introductory Remarks

We have so far only found that particular set of occupation numbers n_i for which the system is in its state of maximum probability. But we really require the *average* numbers n_i *etc.* We can obtain these with the method known as the *Darwin–Fowler method* of mean values.* This method is the most reliable general procedure for deriving statistical distribution functions. The method is also known as the Darwin–Fowler method of *selector variables*, and we apply it here to a closed system described by a canonical ensemble. This consideration when deepened, permits also the rigorous demonstration of the phenomenon of Bose–Einstein condensation, that is, the derivation of the Bose–Einstein distribution under condensation conditions by the method of steepest descents. Exact formal expressions for the partition functions and the mean occupation numbers were first given by Dingle [20] along with the demonstration of Bose–Einstein condensation [20], [21], [17]. As stated by Dingle [20], earlier demonstrations of the phenomenon given in the literature are invalid due to the lack of convergence in the series derived from the method of steepest descents, a problem arising only in the Bose–Einstein case

---

*C.G. Darwin and R.H. Fowler [15]. In the book of F. Reif [59] the Darwin–Fowler method is only referred to in a footnote at the end of Sec. 6.8 and the books of R.H. Fowler [27] and E. Schrödinger [65] are referred to.

and pointed out at about the same time also by Schubert [67] but challenged later by Landsberg [38]. We follow here the considerations of Dingle in his book of 1973 [17] which also comments on the paper of Landsberg.

12.2   The Darwin–Fowler Method of Mean Values

We shall consider only a *system in a heat bath,* and therefore we must introduce the *energy weighting factor* of Chapter 11 (with $E = \sum_i n_i \epsilon_i$ for independent elements), *i.e.*

$$e^{-E/kT}$$

for each arrangement, an arrangement being the distribution of all (say N) elements in packets of $n_1, n_2, n_3, \dots$ among the single particle energy levels $\epsilon_1, \epsilon_2, \epsilon_3, \dots$ with degeneracies $g_1, g_2, g_3, \dots$, taking into account their distinguishability (in classical statistics) or nondistinguishability (in quantum statistics) and the number of elements that can be accommodated on a level, as indicated in Fig. 12.1. We recall from Eqs. (11.22) and (11.40) that

$$F = -kT \ln \sum e^{-E/kT}. \tag{12.1}$$

We set

$$Z = \sum_{\text{arrangements}} e^{-E/kT}. \tag{12.2}$$

This expression Z is called the *canonical partition function* for the system in a heat bath since it evidently expresses a partition of the state of a system into sectors characterized by a specific energy, along with the degeneracy of its states. For independent elements we have

$$E = \sum_i n_i \epsilon_i. \tag{12.3}$$

Therefore (the sum over arrangements being the sum over all eigenenergies)

$$Z = \sum_{\text{arrangements}} e^{-\sum_i n_i \epsilon_i /kT} = \sum_{\text{arrangements}} \prod_i \left(e^{-\epsilon_i/kT} \right)^{n_i}. \tag{12.4}$$

Now set as *single particle energy weighting factor*

$$z_i = e^{-\epsilon_i/kT}. \tag{12.5}$$

Then

$$Z = \sum_{\text{all eigenenergies}} \prod_i z_i^{n_i}, \tag{12.6}$$

where

$$\prod_i z_i^{n_i} = e^{-E/kT},$$

is the overall *energy weighting factor*. Thus we write with the ordering of $\sum$ and $\prod$ left open at this point:

$$Z = \sum \prod_i z_i^{n_i}. \tag{12.7}$$

12.2.1 Mean occupation number $\bar{n}_j$

Z is the sum over all *arrangements of the elements, i.e.* over values of n_i, values of $n_{i+1}, \ldots$, each arrangement being taken with its appropriate energy weighting factor.

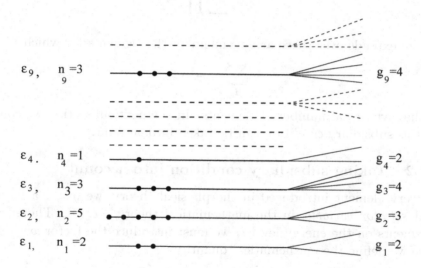

Fig. 12.1 The energy levels ϵ_i with occupation n_i and arbitrarily chosen degeneracies g_i.

Therefore the *mean occupation number, i.e.* the *average over all arrangements*, is

$$\bar{n}_j = \frac{\sum n_j \prod_i z_i^{n_i}}{\sum \prod_i z_i^{n_i}} = \frac{1}{Z} \sum n_j \prod_i z_i^{n_i}. \tag{12.8}$$

Consider the expression (we repeat: $\sum$ meaning the sum over arrangements $E = \sum_i n_i \epsilon_i$, not sum over j)

$$\sum n_j \prod_i z_i^{n_i} = \sum n_j z_j^{n_j} \prod_{i \neq j} z_i^{n_i} = z_j \frac{\partial}{\partial z_j} \left(\sum z_j^{n_j} \right) \prod_{i \neq j} z_i^{n_i}$$

$$= z_j \frac{\partial}{\partial z_j} \sum \prod_i z_i^{n_i} = z_j \frac{\partial}{\partial z_j} Z. \qquad (12.9)$$

It follows that

$$\overline{n}_j = \frac{1}{Z} z_j \frac{\partial}{\partial z_j} Z = z_j \frac{\partial}{\partial z_j} \ln Z, \qquad (12.10)$$

i.e. the mean occupation numbers $\overline{n}_j$ are obtained from the logarithm of the partition function, $\ln Z$. However, the expression in the middle is usually more useful. Both, the thermodynamics and the mean occupation numbers can be calculated from Z. Note that in

$$Z = \sum \prod_i z_i^{n_i}$$

the sum extends over those arrangements of the elements for which

$$\sum_i n_i = N,$$

i.e. the given total number of elements. In the next subsection we consider how this subsidiary condition can be taken into account.

12.2.2   Taking subsidiary condition into account

For every element introduced in the physical theory, we insert a factor ω, called *selector variable*, in the mathematical theory, *i.e.* Z. Therefore for n_i elements on the energy level ϵ_i we must introduce the factor ω^{n_i} into Z. Hence we define the mathematical quantity

$$Z_\omega := \sum \prod_i \omega^{n_i} z_i^{n_i} = \prod_i (\omega z_i)^{n_i}. \qquad (12.11)$$

We now proceed with the theory ignoring the condition $\sum_i n_i = N$. Finally in the answer we pick out the coefficient of ω^N. This will be the result corresponding to that of N elements. Therefore: Z is the coefficient of ω^N in

$$Z_\omega = \sum \prod_i (\omega z_i)^{n_i} \qquad (12.12)$$

(condition $\sum_i n_i = N$ ignored).

12.3 Classical Statistics

We consider classical statistics in two steps, the first step considering the distinguishability of the elements, the second the degeneracy of levels. For reasons of clarity and to ease understanding we write out details in the following.

1. In classical statistics the elements are all *distinguishable*, *i.e.* they all have labels. Therefore the number of ways of dividing the N elements into packets (one packet per energy level, *i.e.* a packet of n_i elements on ϵ_i as in Sec. 4.2, *etc.*) is

$$= \frac{N!}{n_1! n_2! n_3! \cdots} = \frac{N!}{\prod_i n_i!}. \tag{12.13}$$

Therefore in this case

$$Z_\omega = N! \sum_{n_i} \prod_i \frac{(\omega z_i)^{n_i}}{n_i!}, \tag{12.14}$$

where in the sum we no longer have to take account of labelling.

2. In addition, the *degeneracy* has to be taken into account. Any integer value of n_i is allowed in each degenerate level, *i.e.* with $i = 0, 1, 2, \ldots$. The degeneracy of the single particle energy level ϵ_i is g_i. Therefore every contribution i is raised to power g_i and we have (observe the interchange of $\sum$ and $\prod$)

$$\frac{Z_\omega}{N!} = \prod_{i=1}^{\infty} \left(\sum_{n_i=0,1,2,\ldots}^{\infty} \frac{(\omega z_i)^{n_i}}{n_i!} \right)^{g_i}$$

$$= \left(\sum_{n_1=0,1,2,\ldots}^{\infty} \frac{(\omega z_1)^{n_1}}{n_1!} \right)^{g_1} \left(\sum_{n_2=0,1,2,\ldots}^{\infty} \frac{(\omega z_2)^{n_2}}{n_2!} \right)^{g_2}$$

$$\left(\sum_{n_3=0,1,2,\ldots}^{\infty} \frac{(\omega z_3)^{n_3}}{n_3!} \right)^{g_3} \cdots .$$

In more detail this means

$$\frac{Z_\omega}{N!} =$$

$$\left(\underbrace{1}_{n_1=0} + \underbrace{\frac{\omega z_1}{1!}}_{n_1=1} + \underbrace{\frac{(\omega z_1)^2}{2!}}_{n_1=2} + \cdots \right)^{g_1} \left(\underbrace{1}_{n_2=0} + \underbrace{\frac{\omega z_2}{1!}}_{n_2=1} + \underbrace{\frac{(\omega z_2)^2}{2!}}_{n_2=2} + \cdots \right)^{g_2} \cdots ,$$

or perhaps more clearly (the underbracing is explained below)

$$\frac{Z_\omega}{N!} = \left(1 + \underbrace{\frac{\omega z_1}{1!}}_{\cdots} + \frac{(\omega z_1)^2}{2!} + \cdots\right)$$

$$\times \left(1 + \underbrace{\frac{\omega z_1}{1!}}_{\cdots} + \frac{(\omega z_1)^2}{2!} + \cdots\right) \times \left(\cdots\right) \text{ to } g_1 \text{ factors}$$

$$\times \left(1 + \frac{\omega z_2}{1!} + \underbrace{\frac{(\omega z_2)^2}{2!}}_{\cdots} + \cdots\right)$$

$$\times \left(\underbrace{1}_{\cdots} + \frac{\omega z_2}{1!} + \frac{(\omega z_2)^2}{2!} + \cdots\right) \times \left(\cdots\right) \text{ to } g_2 \text{ factors}$$

$$\times \cdots \times \cdots \,. \tag{12.15}$$

In the overall product the term underbraced $\{\cdots\}$ corresponds to: 1 element in the 1st degenerate state of energy level ϵ_1 and 1 element in the 2nd degenerate state of energy level ϵ_1, 2 elements in the 1st degenerate state of energy level ϵ_2, 0 elements in the 2nd degenerate state of energy level ϵ_2, *etc.* All possible arrangements are covered in the full multiplication:

$$\prod_i \left\{\sum_{n_i} \frac{(\omega z_i)^{n_i}}{n_i!}\right\}^{g_i} = \left(e^{\omega z_1}\right)^{g_1} \left(e^{\omega z_2}\right)^{g_2} \cdots = e^{\omega \sum_i g_i z_i}. \tag{12.16}$$

Therefore

$$Z_\omega = N! e^{\omega \sum_i g_i z_i}, \tag{12.17}$$

and hence (see line before Eq. (12.12))

$$Z = \text{coefficient of } \omega^N \text{ in } Z_\omega = N! \frac{(\sum_i g_i z_i)^N}{N!} = \left(\sum_i g_i z_i\right)^N. \tag{12.18}$$

This result does not contain a single approximation! From this result we now obtain the *mean occupation numbers, i.e.*

$$\bar{n}_j \overset{(12.10)}{=} z_j \frac{\partial}{\partial z_j}(\ln Z) = z_j \frac{\partial}{\partial z_j} N \ln\left(\sum_i g_i z_i\right)$$

$$= \frac{N g_j z_j}{\sum_i g_i z_i} = N \frac{g_j e^{-\epsilon_j/kT}}{\sum_i g_i e^{-\epsilon_i/kT}}. \tag{12.19}$$

This result is *exact*. Comparison of this *average* result with the result (4.14) which is the *most probable* value of the occupation number, shows that the right hand sides are identical. Schrödinger[†] remarks:"... it is always attractive and illuminating to see that identically the same result can be obtained by widely different considerations, especially if it is a question of a very general theorem of fundamental importance." The free energy F in this classical statistics is now obtained as

$$F = -kT \ln Z = -NkT \ln \sum_i g_i e^{-\epsilon_i/kT}. \tag{12.20}$$

This result agrees with that of the maximization method, Eq. (5.12).

12.4 Quantum Statistics

We have

$$Z_\omega = \sum \prod_i (\omega z_i)^{n_i}, \quad z_i = e^{-\epsilon_i/kT}, \tag{12.21}$$

where n_i is the number of elements in energy level i. Since elements are indistinguishable according to quantum mechanics, no preliminary calculation of the number of ways of dividing elements into packets $n_1, n_2, \ldots, n_s$ is required. Therefore the sum $\sum$ in quantum statistics refers only to the sum over possible values of n_i.

12.4.1 Fermi–Dirac statistics

In the case of Fermi–Dirac statistics we have

$$n_i = 0 \text{ or } 1 \quad \text{per state.} \tag{12.22}$$

There are g_i states for energy level ϵ_i. Hence we have

$$Z_\omega = \prod_i \sum_{n_i} (\omega z_i)^{n_i}$$
$$= (\underbrace{1}_{\cdots} + \omega z_1)(1 + \underbrace{\omega z_1}_{\cdots})(\underbrace{1}_{\cdots} + \omega z_1) \ldots \text{ to } g_1 \text{ factors}$$
$$\times (1 + \underbrace{\omega z_2}_{\cdots})(1 + \underbrace{\omega z_2}_{\cdots})(1 + \omega z_2) \ldots \text{ to } g_2 \text{ factors}$$
$$\cdots \cdots \tag{12.23}$$

where the contributions underbraced $\{\cdots\}$ correspond to: 0 elements on the first state of level 1, 1 element on the second state of level 1, 0 elements on

---

[†]E. Schrödinger [65], p.27.

the third state of level 1, 1 element on the first state of level 2, 1 element on the second state of level 2, ... We have only 1 and ωz since n is either 0 (first case) or 1 (second case). Therefore

$$Z_\omega = (1 + \omega z_1)^{g_1}(1 + \omega z_2)^{g_2} \cdots = \prod_i (1 + \omega z_i)^{g_i}. \tag{12.24}$$

12.4.2 Bose–Einstein statistics

In the case of Bose–Einstein statistics we have

$$n_i = 0, 1, 2, 3, 4, \ldots, \infty. \tag{12.25}$$

By the same procedure as before it follows that in the present case

$$\begin{aligned} Z_\omega &= (1 + \omega z_1 + (\omega z_1)^2 + (\omega z_1)^3 + \cdots)^{g_1} \\ &\quad \times (1 + \omega z_2 + (\omega z_2)^2 + (\omega z_2)^3 + \cdots)^{g_2} \\ &\quad \cdots\cdots \\ &\quad \cdots\cdots . \end{aligned} \tag{12.26}$$

But

$$1 + \omega z_1 + (\omega z_1)^2 + \cdots = \frac{1}{1 - \omega z_1}.$$

Therefore[‡]

$$Z_\omega = \prod_i (1 - \omega z_i)^{-g_i}. \tag{12.27}$$

Summarizing the two cases we have that Z is the coefficient of ω^N in

$$Z_\omega = \prod_i (1 \pm \omega z_i)^{\pm g_i}, \tag{12.28}$$

where the upper signs apply to Fermi–Dirac statistics, and the lower signs to Bose–Einstein statistics.

12.4.3 Evaluation of the coefficient of $\omega^{\mathbf{N}}$ in $\mathbf{Z_\omega}$

Consider the function

$$\phi(\omega) = a + b\omega + c\omega^2 + \cdots . \tag{12.29}$$

[‡]This expression for $g_i = 1$ (no degeneracy) may be found between Eqs. (3) and (4) of R.B. Dingle [20]. There the method of partial fractions (not applicable in the Fermi–Dirac case) is used to derive Z.

How does one find the coefficient of w^N? We consider the function in the plane of complex w and apply *Cauchy's residue theorem*. We have, to be verified below,

$$\oint \frac{dw}{w} = 2\pi i, \quad \oint \frac{dw}{w^n} = 0 \quad \text{if} \quad n \neq 1. \tag{12.30}$$

In order to verify these results we put

$$w = e^{i\theta}, \quad dw = iw d\theta.$$

Then, with the contour taken along the unit circle around the singular point $w = 0$ as indicated in Fig. 12.2,

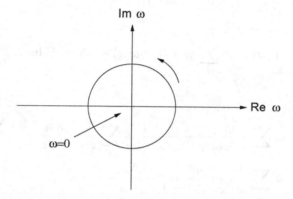

Fig. 12.2 The unit circle contour around $w = 0$ in the complex w-plane.

$$\oint \frac{dw}{w^n} = \int_0^{2\pi} i d\theta e^{i\theta(1-n)} = i \int_0^{2\pi} e^{i(1-n)\theta} d\theta$$

$$= \left[\frac{i e^{i(1-n)\theta}}{i(1-n)} \right]_0^{2\pi} = \frac{e^{i(1-n)2\pi}}{1-n} - \frac{1}{1-n}$$

$$= \frac{1}{1-n} - \frac{1}{1-n} = 0 \quad \text{provided } n \neq 1.$$

For $n = 1$ the first of Eqs. (12.30) is seen to follow immediately from the equation in the first line. It follows that the coefficient of w^N in $\phi(w)$ is

$$= \frac{1}{2\pi i} \oint \frac{\phi(w) dw}{w^{N+1}}. \tag{12.31}$$

Hence the only surviving term is given by the important relation:

$$\frac{1}{2\pi i} \oint \frac{(\text{coefficient of } w^N) w^N dw}{w^{N+1}} = \text{coefficient of } w^N \text{ in } \phi(w). \tag{12.32}$$

Therefore in view of Eq. (12.28)

$$Z = \frac{1}{2\pi i} \oint \frac{Z_\omega}{\omega^{N+1}} d\omega = \frac{1}{2\pi i} \oint \frac{1}{\omega^{N+1}} \left\{ \prod_i (1 \pm \omega z_i)^{\pm g_i} \right\} d\omega. \qquad (12.33)$$

In the following and later considerations one should keep in mind *Cauchy's theorem* which says that the value of an integral is unaffected by deformations of the contour in the complex plane, provided no singularity of the integrand is crossed thereby. Writing Z as

$$Z = \frac{1}{2\pi i} \oint e^{f(\omega)} d\omega, \qquad (12.34)$$

we have

$$f(\omega) = \pm \sum_i g_i \ln(1 \pm \omega z_i) - (N+1) \ln \omega. \qquad (12.35)$$

Then

$$\begin{aligned}
f'(\omega) &= \sum_i \frac{g_i z_i}{1 \pm \omega z_i} - \frac{N+1}{\omega} = \sum_i \frac{1}{\omega} \frac{g_i z_i}{\omega \omega^{-1} \pm z_i} - \frac{N+1}{\omega} \\
&= \sum_i \frac{1}{\omega} \frac{g_i}{(\omega z_i)^{-1} \pm 1} - \frac{N+1}{\omega} \\
&= \frac{1}{\omega} \left[\sum_i \frac{g_i}{(\omega z_i)^{-1} \pm 1} - (N+1) \right].
\end{aligned} \qquad (12.36)$$

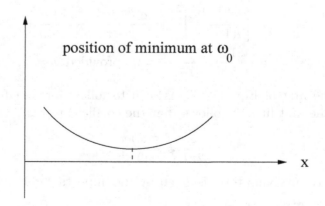

Fig. 12.3 The position of the minimum at ω_0 when travelling along the real axis.

Also (the second step is written to show the cancellation):

$$f''(\omega) = -\frac{1}{\omega^2}\left[\sum_i \frac{g_i}{(\omega z_i)^{-1} \pm 1} - (N+1)\right]$$

$$-\frac{1}{\omega}\sum_i \frac{g_i}{[(\omega z_i)^{-1} \pm 1]^2}\frac{d}{d\omega}\left(\frac{1}{\omega z_i}\right)$$

$$= -\frac{1}{\omega^2}\left[\sum_i \frac{g_i}{(\omega z_i)^{-1} \pm 1} - (N+1)\right]$$

$$+\frac{1}{\omega^2}\sum_i \frac{g_i\{(\omega z_i)^{-1} \pm 1 \mp 1\}}{[(\omega z_i)^{-1} \pm 1]^2}$$

$$= \frac{(N+1)}{\omega^2} \mp \frac{1}{\omega^2}\sum_i \frac{g_i}{[(\omega z_i)^{-1} \pm 1]^2}. \qquad (12.37)$$

Equation (12.36) gives a minimum for $\omega = \omega_0$, *i.e.* we have at ω_0:

$$f'(\omega_0) = 0, \quad \text{or} \quad N+1 = \sum_i \frac{g_i}{(\omega_0 z_i)^{-1} \pm 1}. \qquad (12.38)$$

If $f(\omega)$ is analytic (*i.e.* regular) in the ω-plane, it obeys Laplace's equation

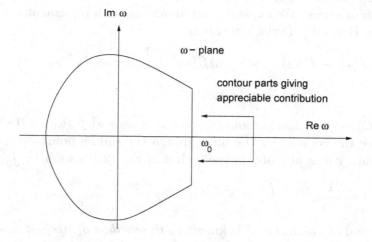

Fig. 12.4 Contour parts giving appreciable contribution.

in 2 dimensions, *i.e.*[§]

$$\frac{\partial^2 f}{\partial x^2} + \frac{\partial^2 f}{\partial y^2} = 0. \qquad (12.39)$$

[§]Such a function (which cannot have a maximum or minimum) satisfying the two-dimensional Laplace equation is known as a *potential function*, *cf.* H.J.W. Müller–Kirsten [50], Sec. 3.7.

Therefore

$$\frac{\partial^2 f}{\partial x^2} = -\frac{\partial^2 f}{\partial y^2}. \tag{12.40}$$

Hence from curvature — actually from the *curvature formula* — if we have a stationary point of $f(\omega)$ with a minimum at $\omega = \omega_0$ in one direction (along the real axis as indicated in Fig. 12.3), we have a maximum at $\omega = \omega_0$ in the orthogonal direction, *i.e.* along the path parallel to the imaginary axis, as indicated in Fig. 12.4. This is the reason why such stationary points are described as *saddle points*. We have in the present case

$$Z = \frac{1}{2\pi i} \oint e^{f(\omega)} d\omega, \tag{12.41}$$

where the integral is understood as a contour integral in the complex ω-plane taken in an anti-clockwise sense along a closed path encircling the singularity at $\omega = 0$. It is this contour which we shall deform later to pass through a stationary point (saddle point) and thus to enable us to extract the dominant contribution to the integral. As Dingle [17] (p.134) emphasizes, the asymptotic nature of the final expansion really originates through extending the range of integration beyond a circle of convergence whose radius in the plane of $F = -f$ is fixed by the next zero of the derivative of F; a mere change in direction of the path crossing this fateful circle cannot materially affect the resultant series. We expand $f(\omega)$ about ω_0, with increment $\omega - \omega_0 = iy$, $d\omega = idy$. Hence by Taylor's theorem

$$f(\omega) = f(\omega_0) + (\omega - \omega_0)f'(\omega_0) + \frac{1}{2}(\omega - \omega_0)^2 f''(\omega_0) + \cdots$$

$$= f(\omega_0) - \frac{1}{2}y^2 f''(\omega_0) + \cdots, \tag{12.42}$$

since $f'(\omega_0) = 0$ at the stationary point $\omega = \omega_0$ and $f''(\omega_0) > 0$. Therefore we replace the contour by the line through the saddle point and parallel to the imaginary axis and obtain (note that in Eq. (2.20) we have $\int_0^\infty$):

$$Z = \frac{1}{2\pi i}e^{f(\omega_0)} \int_{-\infty}^{\infty} e^{-y^2 f''(\omega_0)/2} idy \stackrel{(2.20)}{=} \frac{e^{f(\omega_0)}}{\sqrt{2\pi f''(\omega_0)}}. \tag{12.43}$$

This method of evaluating Z is known as the *method of steepest descents* and can be traced back to the work of Riemann and in a related sense to earlier work of Laplace (*cf.* Dingle [17], p.133).[¶] We have $f'(\omega_0) = 0$. Therefore we have (*cf.* Eq. (12.36))

$$N \underbrace{[+1]}_{\text{negligible}} = \sum_i \frac{g_i}{(\omega_0 z_i)^{-1} \pm 1}. \tag{12.44}$$

[¶]The modern application is due to P. Debye [16]. See also B.L. Van der Waerden [81].

This equation determines the value of ω_0. We will see in a moment that this equation corresponds to the condition

$$N = \sum_i \overline{n}_i. \tag{12.45}$$

From Eq. (12.43) we obtain (as earlier defined by Eq. (7.28), there with $\epsilon_0 = kT \ln \omega_0$, ω_0 corresponding to the present ω_0, and as remarked in connection with Eq. (11.33))

$$
\begin{aligned}
\ln Z \; &= \; f(\omega_0) - \frac{1}{2} \ln[2\pi f''(\omega_0)] \\
&\overset{(12.35)}{=} \; \pm \sum_i g_i \ln(1 \pm \omega_0 z_i) - (N+1) \ln \omega_0 - \frac{1}{2} \ln[2\pi f''(\omega_0)] \\
&\equiv \; \ln Z(z_i, \omega_0(z_i)).
\end{aligned} \tag{12.46}
$$

Here (see Example 12.10)

$$g_i \ln(1 \pm \omega_0 z_i) \propto V,$$

where V is the volume of the system, and

$$\frac{1}{2} \ln[2\pi f''(\omega_0)] \propto \ln V,$$

which is therefore normally negligible for large systems. Then

$$\frac{\partial}{\partial z_j} \ln Z \simeq \pm \sum_i g_i \left(\frac{\pm \omega_0}{1 \pm \omega_0 z_i} \right) \delta_{ij} = \frac{g_j \omega_0}{1 \pm \omega_0 z_j}, \tag{12.47}$$

and, with Eqs. (12.10) and (12.5),

$$
\begin{aligned}
\overline{n}_j &= z_j \frac{d}{dz_j} \ln Z = z_j \left(\frac{\partial \ln Z}{\partial z_j} + \frac{\partial \ln Z}{\partial \omega_0} \frac{\partial \omega_0}{\partial z_j} \right) \\
&= \frac{g_j \omega_0 z_j}{1 \pm \omega_0 z_j} = \frac{g_j}{(\omega_0 z_j)^{-1} \pm 1} \\
&= \frac{g_j}{\omega_0^{-1} e^{\epsilon_j/kT} \pm 1},
\end{aligned} \tag{12.48}
$$

with $\partial \ln Z/\partial \omega_0 = 0$ in this approximation in view of Eq. (12.38). The expression (12.48) gives the *mean number of elements* (particles, atoms) of the total of N in the volume V which occupy at temperature T the 1-particle energy level ϵ_j. For the derivation to be reliable one should check that higher order contributions (those involving higher order derivatives of $f(\omega)$) are initially decreasing in magnitude, so that the expansion about the saddle point does indeed yield an asymptotic expansion. Unfortunately in the case of Bose–Einstein statistics in the approach to $T = 0$, under conditions for the phenomenon of Bose–Einstein condensation to occur, this is not the case, and these higher order contributions are of comparable magnitude as the leading contribution, so that the result Eq. (12.48) cannot be trusted in that domain, and with the above calculation. There is, however, a way to modify the calculation and hence to save the result. This calculation will be explained along with the phenomenon of Bose–Einstein condensation in Sec. 12.5.

12.5   Bose–Einstein Condensation

Finally we consider briefly the phenomenon of *Bose–Einstein condensation* which has attracted considerable attention in recent years since its occurrence has now been definitely established. We consider first how the condensation is understood theoretically on the basis of the Bose–Einstein distribution, and then — since as mentioned earlier most derivations are invalid in the condensation domain — we repeat the derivation of this distribution under condensation conditions following the rigorous arguments of Dingle.[||]

12.5.1   The phenomenon of Bose–Einstein condensation

We consider the *Bose–Einstein distribution* (12.48),[**] *i.e.*

$$\bar{n}_j = \frac{g_j}{\omega_0^{-1} e^{\epsilon_j/kT} - 1} = \frac{g_j}{e^{(\epsilon_j - \mu)/kT} - 1},\tag{12.49}$$

where $\omega_0 = e^{\mu/kT}$. For a value of μ such that $\mu < \epsilon_j$, all j, the mean occupation number $\bar{n}_j$ decreases with $T \to 0$. However, this implies also that

$$N = \sum_j \bar{n}_j$$

[||]R.B. Dingle [17], pp.267-271. Another rigorous derivation has been given by S. Greenspoon and R.K. Pathria [29], C.S. Zasada and R.K. Pathria [86].
[**]See also T.L. Hill [34], pp.439-440.

decreases to zero in that limit. Now assume the ground state energy is $\epsilon_0{}^{\dagger\dagger}$ and that $\epsilon_0 - \mu = \delta$, where δ is infinitesimal. In this case $\mu = \epsilon_0 - \delta$, and

$$\text{for} \quad T \to 0 : \qquad \overline{n}_{j\neq 0} = \frac{g_{j\neq 0}}{e^{(\epsilon_{j\neq 0}-\epsilon_0+\delta)/kT} - 1} \to 0,$$

$$\text{but} \qquad\qquad \overline{n}_0 = \frac{g_0}{e^{(\epsilon_0-\mu)/kT} - 1} \to g_0 \frac{kT}{\epsilon_0 - \mu}. \qquad (12.50)$$

Since

$$N = \sum_j \overline{n}_j \overset{T\to 0}{\longrightarrow} \overline{n}_0, \qquad (12.51)$$

we can identify $g_0 kT/(\epsilon_0 - \mu)$ with N, *i.e.*

$$\epsilon_0 - \mu = g_0 \frac{kT}{N}, \qquad (12.52)$$

so that with $T \to 0$, $\mu \to \epsilon_0$. This means, in the limit $T \to 0$ the only thermodynamically sensible value of μ is $\mu = \epsilon_0$. We also see that in this limit all particles populate the lowest one-particle state, all higher states being empty. This phenomenon is described as *Bose–Einstein condensation*. That the phenomenon of Bose–Einstein condensation actually occurs was confirmed in 1995 in dilute gases of alkali atoms by E.A. Cornell and C.E. Wieman of the Joint Institute for Laboratory Astrophysics in Boulder and by W. Ketterle of the MIT in Cambridge, Mass.; for this achievement they were awarded the Nobel Prize in 2001.[‡‡]

12.5.2 Derivation of the Bose–Einstein distribution function under condensation conditions

As already stated, we now rederive the Bose–Einstein distribution following the arguments of Dingle.[§§] We start from Eq. (12.33), *i.e.* the Bose–Einstein partition function

$$Z = \frac{1}{2\pi i} \oint \frac{d\omega}{\omega^{N+1} \prod_i (1 - \omega z_i)^{g_i}}. \qquad (12.53)$$

This result of the *Darwin–Fowler method* is an *exact expression*. For convenience of the present considerations we assume the states to be nondegenerate, *i.e.* $g_i = 1$. As before we consider the exponent $f(\omega)$ of the

---

[††]Throughout the entire text we use the symbol ϵ_0 with different meanings (Fermi energy, ground state energy, chemical potential); we assume the relevant meaning is clear from the respective context.

[‡‡]For a review see K. Burnett, M. Edwards and C.W. Clark [9].

[§§]R.B. Dingle [17], pp.267-271.

exponentiated integrand $e^{f(\omega)}$,

$$f(\omega) = -(N+1)\ln\omega - \sum_i \ln(1 - \omega z_i) \equiv -\overline{F}(\omega), \quad g_i = 1. \qquad (12.54)$$

The *stationary point* of this function is at $\omega = \omega_0(z_i)$ where (a subscript $j, j \geq 1$, meaning j-th derivative)

$$0 = \omega_0 \overline{F}_1(\omega_0) = N + 1 - \sum_i \frac{\omega_0 z_i}{1 - \omega_0 z_i}. \qquad (12.55)$$

Following Dingle [17] we want to consider the usual saddle-point method — but now specifically under condensation conditions. Under the usual non-condensation conditions the procedure poses no problem — and for the purpose of demonstrating the need for a reconsideration, we reconsider this first. To this end we consider higher order derivatives of the function $f(\omega) \equiv -\overline{F}(\omega)$ and their contributions to the partition function Z, or better, to $\ln Z$. The higher order derivatives are readily found to be given by the following expression at the stationary point $\omega_0 = \omega_0(z_i)$:

$$\overline{F}_j(\omega_0) \equiv \left[\frac{d^j}{d\omega^j}\overline{F}(\omega)\right]_{\omega_0} = -\frac{(j-1)!}{\omega_0^j}\left[(-1)^j(N+1) + \sum_i \left(\frac{\omega_0 z_i}{1 - \omega_0 z_i}\right)^j\right]. \qquad (12.56)$$

We observe that the second derivative is given by:

$$\left[\frac{d^2\overline{F}}{d\omega^2}\right]_{\omega_0} = -\frac{1}{\omega_0^2}\left[(N+1) + \sum_i \left(\frac{\omega_0 z_i}{1 - \omega_0 z_i}\right)^2\right] < 0, \qquad (12.57)$$

as required for a maximum of $\overline{F}$ (minimum of the integrand) for a path along the real axis, so that the integration contour is to be taken through the maximum of the integrand and parallel to the imaginary ω-axis. We obtain from Eq. (12.46) for the present case of Bose–Einstein statistics and at the stationary point ω_0 (for convenience with degeneracies $g_i = 1$):

$$\ln Z = -\sum_i \ln(1 - \omega_0 z_i) - (N+1)\ln\omega_0 - \frac{1}{2}\ln[2\pi f''(\omega_0)] + \cdots$$

$$= -\frac{1}{2}\ln 2\pi - (N+1)\ln\omega_0 - \sum_i \ln(1 - \omega_0 z_i)$$

$$-\frac{1}{2}\ln\left[\frac{1}{\omega_0^2}\left\{(N+1) + \sum_i \left(\frac{\omega_0 z_i}{1 - \omega_0 z_i}\right)^2\right\}\right]. \qquad (12.58)$$

Here from the subsidiary condition (12.55):

$$N + 1 = \sum_i \frac{\omega_0 z_i}{1 - \omega_0 z_i} \left(\frac{1 - \omega_0 z_i}{1 - \omega_0 z_i} \right), \tag{12.59}$$

so that

$$\ln Z = -\frac{1}{2} \ln 2\pi - (N+1) \ln \omega_0 - \sum_i \ln(1 - \omega_0 z_i)$$
$$- \frac{1}{2} \ln \left[\frac{1}{\omega_0} \sum_i \frac{z_i}{(1 - \omega_0 z_i)^2} \right] - \cdots . \tag{12.60}$$

In Example 12.10, which considers normal conditions, *i.e.* not those relevant for condensation, the contribution of the term containing the second order derivative (the fourth term in Eq. (12.60)) is shown to be of order $\ln V$ compared with the preceding term which is of order V, and hence this term with second order derivative, and later terms can be dropped. However, since $\epsilon_i > \epsilon_{i-1}$, and therefore $\bar{n}_i < \bar{n}_{i-1}$, the lowest single particle level has the greatest equilibrium occupation[¶¶] and hence in this last case[***]

$$(\omega_0 z_0)^{-1} = 1 + (\bar{n}_0)^{-1} \tag{12.61}$$

has the smallest deviation from unity. Looking more closely at the fourth contribution in Eq. (12.60), we see that its dominant contribution is equal and opposite in sign to the corresponding contribution contained in the third term since:

$$-\frac{1}{2} \ln \left[\frac{\omega_0^{-1} z_0}{(1 - \omega_0 z_0)^2} + \cdots \right] \simeq -\frac{1}{2} \ln(\omega_0^{-1} z_0) - \frac{1}{2} \ln \frac{1}{(1 - \omega_0 z_0)^2} + \cdots$$
$$= \ln(1 - \omega_0 z_0) + \cdots . \tag{12.62}$$

A detailed look at the higher order terms reveals that these also contain contributions of comparable magnitude, so that there is no decrease from the leading term, and the expansion of steepest descents becomes meaningless. One might think that the problem arises owing to the factor $(1 - \omega_0 z_0)$ in the integrand of Z, Eq. (12.53), and if this factor were separated off from the integrand before the stationary point is determined, the method could be saved. But this is not the case since then the next factor in the product shifts the expansion point and then recreates the problem. Thus the usual steepest descents derivation of the Bose–Einstein distribution is faulty.

[¶¶]This follows from the Bose–Einstein distribution (12.49): An ϵ_j larger in the denominator on the right implies a smaller $\bar{n}_j$ in the numerator on the left.
[***]From Eq. (12.49) for $j = 0$ we obtain: $\bar{n}_0 = [(\omega_0 z_0)^{-1} - 1]^{-1}$ for $g_0 = 1$.

The rigorous method of uniform asymptotic expansions of Dingle [17] ('uniform' meaning here based on higher transcendental functions, see Dingle [17], p.217) does not move the expansion point when a singular factor is singled out like the factor $1/(1-wz_0)$ here. This simple pole at $w \to w_0 = z_0^{-1}$ is constructed to be a certain distance (below called u_0) away from a new stationary point (below at $u = 0$). One now considers the partition function Z in the following form and with change of variable of integration explained below:

$$Z \equiv \frac{1}{2\pi i} \oint \frac{dw}{w^{N+1} \prod_i (1 - wz_i)} = \frac{1}{2\pi i} \oint e^{-\overline{F}(w)} G(w) dw$$

$$= \frac{1}{2\pi i} \left(-\frac{1}{z_0} \right) \int_{s.p.} e^{-\overline{F}(u)} G(u) du. \qquad (12.63)$$

Here s.p. stands for 'stationary point' (s.p. method of derivation of the asymptotic expansion from this integral representation — see Dingle [17], p.131). The function $\overline{F}(u)$ is chosen such that it contains the dominant pole at $u = u_0$, i.e., with $z_0^{-1} - w = u_0 - u$,

$$\left. \begin{array}{l} e^{-\overline{F}(w)} = \frac{1}{(1-wz_0)} e^{-F(w)}, \quad \overline{F}(w) = F(w) + \ln(1 - wz_0), \\[2mm] e^{-\overline{F}(u)} = \frac{1}{(u-u_0)} e^{-F(u)}, \quad \overline{F}(u) = F(u) + \ln(u - u_0). \end{array} \right\} \qquad (12.64)$$

The stationary (or critical) point is set at $u = 0$, with

$$\left. \frac{d\overline{F}(u)}{du} \right|_{u=0} = 0,$$

making $\overline{F}(u)$ quadratic at the stationary point (i.e. not $F(u)$), and this is maintained there and taken as the point of expansion. Then at $u = 0$:

$$\overline{F}_1(0) = F_1(0) - \frac{1}{u_0} = 0,$$

$$\overline{F}_j(0) = \left. \frac{d^j \overline{F}(u)}{du^j} \right|_{u=0} = \left. \frac{d^j F(u)}{du^j} \right|_{u=0} - \frac{(j-1)!}{u_0^j}, \qquad j \geq 1. \quad (12.65)$$

For sufficiently small $|u_0|$ these derivatives yield large contributions to the integral stemming from the logarithmic part in $\overline{F}(u)$. The idea is to avoid these large contributions by a method which relates the leading part of the integral to the integral representation of a known special function, and this means that which retains only terms up to the second derivative (with $|F_2|$ large, otherwise the uniform expansion is pointless, Dingle [17], p.220) leaving the rest in the remaining part of the integrand but with derivatives $F_3, F_4, \ldots$ of $F(u)$. Expanding $F(u)$ about $u = 0$,

$$F(u) = F_0 + F_1 u + \sum_{i=2}^{\infty} F_i \frac{u^i}{i!}, \quad F(0) = F_0,$$

so that

$$\overline{F}(u) = F_0 + F_1 u + \sum_{i=2}^{\infty} F_i \frac{u^i}{i!} + \ln(u - u_0), \tag{12.66}$$

and demanding $\overline{F}_1(0) = 0$, we obtain

$$F_1 + \frac{1}{-u_0} = 0, \quad F_1 = \frac{1}{u_0}. \tag{12.67}$$

Then (grouping factors conveniently)

$$\int_{\text{s.p.}} e^{-\overline{F}(u)} G(u) du = \int_{\text{s.p.}} e^{-F_1 u - F_2 u^2/2} e^{-F_0 - \sum_{i=3}^{\infty} F_i u^i/i! - \ln(u - u_0)} G(u) du$$

$$= \int_{\text{s.p.}} \frac{du\, G(u)}{(u - u_0)} e^{-F_1 u - F_2 u^2/2} e^{-F_0 - \sum_{i=3}^{\infty} F_i u^i/i!}$$

$$= \int_{\text{s.p.}} \frac{du\, G(u)}{(u - u_0)} e^{-F_1 u - F_2 u^2/2} e^{-F(u) + F_1 u + F_2 u^2/2}. \tag{12.68}$$

Taking now as expansion point the stationary point $u = 0$ of $\overline{F}(u)$, we obtain (later, for our purposes here, we shall require only leading terms)

$$\int_{\text{s.p.}} e^{-\overline{F}(u)} G(u) du = \sum_{r=0}^{\infty} \left\{ \text{coefficient of } u^r \text{ in } G(u) e^{-F(u) + F_1 u + F_2 u^2/2} \right\}$$

$$\times \left\{ \int_{\text{s.p.}} \frac{du}{(u - u_0)} e^{-F_1 u - F_2 u^2/2} u^r \right\}. \tag{12.69}$$

This expansion now contains instead of the derivatives $\overline{F}_j(0)$, the derivatives of $F(u)$. In order to be able to evaluate the integral on the right, the factor u^r is expanded by the binomial theorem in powers of

$$v := (u - u_0)\sqrt{F_2}, \quad u = \frac{v}{\sqrt{F_2}} + u_0. \tag{12.70}$$

Thus

$$u^r = \frac{1}{(\sqrt{F_2})^r}[(u - u_0)\sqrt{F_2} + u_0\sqrt{F_2}]^r$$

$$= \sum_{s=0}^{r}\binom{r}{s}(u_0\sqrt{F_2})^{r-s}(u - u_0)^s\sqrt{F_2}^{s-r}$$

$$= \sum_{s=0}^{r}\binom{r}{s}(u_0\sqrt{F_2})^{r-s}v^s(\sqrt{F_2})^{-r}. \tag{12.71}$$

Then the integral in Eq. (12.69) becomes

$$\int_{\text{s.p.}}\frac{du}{(u - u_0)}e^{-F_1 u - F_2 u^2/2}u^r$$

$$= \int_{\text{s.p.}}e^{-F_1(v/\sqrt{F_2} + u_0) - F_2(v/\sqrt{F_2} + u_0)^2/2} \times \frac{\sqrt{F_2}}{v}$$

$$\times \sum_{s=0}^{r}\binom{r}{s}(u_0\sqrt{F_2})^{r-s}v^s(\sqrt{F_2})^{-r}\frac{dv}{\sqrt{F_2}},$$

and with the help of Eq. (12.67) this becomes

$$\int_{\text{s.p.}}\frac{du}{(u - u_0)}e^{-F_1 u - F_2 u^2/2}u^r$$

$$\overset{(12.67)}{=}e^{-1 - F_2 u_0^2/2}F_2^{-r/2} \times \sum_{s=0}^{r}\binom{r}{s}(u_0\sqrt{F_2})^{r-s}$$

$$\times \int_{\text{s.p.}}e^{-v^2/2 - v(\sqrt{F_2}u_0 + 1/u_0\sqrt{F_2})}v^{s-1}dv. \tag{12.72}$$

The stationary point of $\overline{F}(u)$ was constructed to be at $u = 0$, i.e. $v = -u_0\sqrt{F_2}$.

The evaluation of the integral is now achieved by comparison with a known integral representation of a special function. Such representations are looked up in relevant texts on special functions or in Tables of Integrals. In the present case the relevant integral representation is that of the parabolic cylinder function $D_\nu(\mu)$ (familiar from the study of the quantum mechanics of harmonic oscillators, there mostly expressed in terms of the related Hermite polynomials), i.e.[†††]

$$D_\nu(\mu) = \frac{1}{\sqrt{2\pi}}i^{\mp\nu}e^{\mu^2/4}\int_{-\infty}^{\infty}e^{\pm i\mu v - v^2/2}v^\nu dv, \qquad -1 = e^{\pm i\pi}. \tag{12.73}$$

[†††]See e.g. W. Magnus and F. Oberhettinger [40], p.92.

Also one sets, with F_0 now included in the argument of the exponential (Dingle [17], p.219, G_i like F_i denoting the i-th derivative at (here) $u = 0$, see Dingle [17], p.114):

$$\mathcal{U}_r = r! \times \text{coefficient of } u^r \text{ in } G(u)e^{-F(u)+F_0+F_1u+F_2u^2/2},$$
$$\mathcal{U}_0 = G_0, \quad \mathcal{U}_1 = G_1, \quad \mathcal{U}_2 = G_2, \quad \mathcal{U}_3 = G_3 - G_0F_3, \dots . \qquad (12.74)$$

Then with $(\sqrt{F_2} \equiv -i\sqrt{-F_2})$

$$\mu := u_0\sqrt{-F_2} = iu_0\sqrt{F_2}, \quad u_0 > 0,$$
$$\bar{\mu} := \mu - \frac{1}{\mu},$$
$$i\bar{\mu} = i\left(\mu - \frac{1}{\mu}\right) = -\left(u_0\sqrt{F_2} + \frac{1}{u_0\sqrt{F_2}}\right), \qquad (12.75)$$

(implying in any case $F_2 \neq 0$, if not very large, and $\bar{\mu}$ large and negative) we have for the integral in Eq. (12.72)

$$\int_{\text{s.p.}} e^{-v^2/2 - v(\sqrt{F_2}u_0 + 1/u_0\sqrt{F_2})}v^{s-1}dv$$
$$= \int_{-\infty}^{\infty} e^{i\bar{\mu}v - v^2/2}v^{s-1}dv$$
$$= \sqrt{2\pi}i^{s-1}e^{-\bar{\mu}^2/4}D_{s-1}(\bar{\mu}). \qquad (12.76)$$

Evaluating the integral $\int_{\text{s.p.}}$ in Eq. (12.72) by comparison with the upper sign integral in Eq. (12.73), we have to identify $\bar{\mu}$ here in Eq. (12.76) with μ in Eq. (12.73), which means the position of the stationary point of $\overline{F}(u)$ at $v = -u_0\sqrt{F_2} = +iu_0\sqrt{-F_2}$, and therefore for $u_0 > 0$ the stationary point at $v = -u_0\sqrt{F_2}$ must be regarded as possessing a positive imaginary part for the path of integration through the stationary point from $\Re v = -\infty$ to ∞. Thus for the entire expression of Eq. (12.72) we obtain:

$$\int_{\text{s.p.}} \frac{du}{(u - u_0)}e^{-F_1u - F_2u^2/2}u^r = e^{-1-F_2u_0^2/2}F_2^{-r/2}\sum_{s=0}^{r}\binom{r}{s}(u_0\sqrt{F_2})^{r-s}$$
$$\times \sqrt{2\pi}i^{s-1}e^{-\bar{\mu}^2/4}D_{s-1}(\bar{\mu}), \qquad (12.77)$$

and therefore from Eq. (12.69) we obtain (with $(-i\mu)^s = (u_0\sqrt{F_2})^s$, $\mu^2 = -F_2 u_0^2$ (Eq. (12.75)) and noting that the binomial coefficient contains $r!$)

$$\int_{\text{s.p.}} e^{-\overline{F}(u)} G(u)\,du = \sum_{r=0}^{\infty} \left\{ \text{coefficient of } u^r \text{ in } G(u)e^{-F(u)+F_1 u+F_2 u^2/2} \right\}$$

$$\times e^{-1-F_2 u_0^2/2} F_2^{-r/2} \sum_{s=0}^{r} \binom{r}{s} (u_0\sqrt{F_2})^{r-s}$$

$$\times \sqrt{2\pi}\, i^{s-1} e^{-\overline{\mu}^2/4} D_{s-1}(\overline{\mu})$$

$$= \sqrt{2\pi}\, e^{-F_0+\mu^2/2-\overline{\mu}^2/4-1} \sum_{r=0}^{\infty} U_r u_0^r$$

$$\times \sum_{s=0}^{r} \frac{1}{s!(r-s)!} D_{s-1}(\overline{\mu}) \frac{i^{s-1}}{(-i\mu)^s}, \tag{12.78}$$

and then again with $\sqrt{-F_2} = (u_0/\mu)^{-1}$, we obtain

$$\int_{\text{s.p.}} e^{-\overline{F}(u)} G(u)\,du = \sqrt{\frac{2\pi}{F_2}}\, e^{-F_0+\mu^2/2-\overline{\mu}^2/4-1} \sum_{r=0}^{\infty} U_r u_0^r \sum_{s=0}^{r} \frac{1}{s!(r-s)!}$$

$$\times D_{s-1}(\overline{\mu}) \frac{1}{(-\mu)^s} \left(-\frac{u_0}{\mu}\right)^{-1}, \tag{12.79}$$

which is the result Eq. (34), p.265, of Dingle [17]. Note that the path of integration through the stationary point is indicated by the integration from $-\infty$ to ∞ in Eq. (12.73). When $\overline{\mu}$ is large and negative, the function $D_{s-1}(\overline{\mu})$ can be approximated by the leading term of its asymptotic expansion for negative variable. But here we do not require the general case, so we specialize immediately to $s=0$ and obtain $D_{-1}(\overline{\mu})$ from the following integral representation (see *e.g.* Dingle [17], Eq. (35), p.265, or Magnus and Oberhettinger [40], p.93)

$$p!\,D_{-1-p}(\overline{\mu}) = e^{-\overline{\mu}^2/4} \int_0^{\infty} e^{-\overline{\mu}v-v^2/2} v^p\,dv, \tag{12.80}$$

so that, with $p=0$ and a change of variable to $x = v + \overline{\mu}$,

$$D_{-1}(\overline{\mu}) = e^{\overline{\mu}^2/4} \int_0^{\infty} e^{-(v+\overline{\mu})^2/2}\,dv = e^{\overline{\mu}^2/4} \int_{\overline{\mu}}^{\infty} e^{-x^2/2}\,dx$$

$$= e^{\overline{\mu}^2/4} \left[\int_0^{\infty} e^{-x^2/2}\,dx - \int_0^{\overline{\mu}} e^{-x^2/2}\,dx \right]$$

$$= e^{\overline{\mu}^2/4} \left[\frac{1}{2}\sqrt{2\pi} - \sqrt{\frac{\pi}{2}} \phi\left(\frac{\overline{\mu}}{\sqrt{2}}\right) \right], \tag{12.81}$$

where ϕ is the *error function*. Inserting the leading term of this result into Eq. (12.79) for $r = s = 0$, we obtain

$$\int_{\text{s.p.}} e^{-F(u)} G(u)\, du = \sqrt{\frac{2\pi}{F_2}} e^{-F_0 + \mu^2/2 - 1} \frac{\sqrt{2\pi}}{2} u_0 \left(-\frac{u_0}{\mu} \right)^{-1}. \qquad (12.82)$$

Before we continue with the calculation, we consider how and where the *condensation condition* comes in. We observed in Sec. 12.5.2 after Eq. (12.60) that the one-particle ground state energy ϵ_0 has the greatest equilibrium occupation $\bar{n}_0$, and the distribution function $\bar{n}_0$ at normal temperatures increases with a lowering of the temperature T. We write now in order to take explicitly into account the deviation of the factor $(1 - wz_0)$ from its value zero at the original saddle point value w_0 of w (*cf.* Eq. (12.61)):

$$(w_0 z_0)^{-1} = 1 + \bar{n}_0^{-1}, \qquad w_0^{-1} = (1 + \bar{n}_0^{-1}) z_0, \qquad (12.83)$$

and

$$w_0 = \frac{1}{z_0(1 + \bar{n}_0^{-1})}, \qquad \bar{n}_0 \gg 1. \qquad (12.84)$$

Thus the quantity u_0 defined above Eq. (12.64) as the distance of the new saddle point at $u = 0$ away from this point is given by

$$u_0 = z_0^{-1} - w_0 = z_0^{-1} - \frac{1}{z_0(1 + \bar{n}_0^{-1})} = z_0^{-1}\left[1 - \frac{1}{1 + \bar{n}_0^{-1}} \right]$$

$$= \frac{1}{z_0} \frac{\bar{n}_0^{-1}}{1 + \bar{n}_0^{-1}} \simeq \frac{1}{z_0} \bar{n}_0^{-1} = \frac{1}{z_0 \bar{n}_0}, \qquad (12.85)$$

with $\bar{n}_0 \gg 1, \bar{n}_0^{-1} \ll 1$. Also, the distance u_0 is very small under condensation conditions. Thus, since $\sqrt{F_2} \neq 0$, we deduce from Eq. (12.75) that $\bar{\mu} \simeq -1/\mu$ is large and negative, and μ is small.

We have from Eqs. (12.54) and (12.64):

$$F(w) = (N + 1) \ln w + \sum_{i \neq 0} \ln(1 - w z_i). \qquad (12.86)$$

Now Z of Eq. (12.63) is for $g_i = 1$ (and recall from Eq. (12.66) that F_0, F_2 are the values at the saddle point $u = 0$):

$$Z = \frac{1}{2\pi i}\left(-\frac{1}{z_0} \right) \int_{\text{s.p.}} e^{-F(u)} G(u)\, du, \qquad (12.87)$$

and hence we obtain (for the second line see after Eq. (12.88) below)

$$Z \overset{(12.82)}{=} \frac{1}{2\pi i}\left(-\frac{1}{z_0}\right)\frac{2\pi}{2\sqrt{F_2}}e^{-F_0+\mu^2/2-1}\mathcal{U}_0\left(-\frac{\mu}{u_0}\right)$$

$$\overset{(12.86)}{=} \frac{\mu}{2u_0 z_0\sqrt{-F_2}}e^{\mu^2/2-1}e^{-(N+1)\ln w_0 - \sum_{i\neq0}\ln(1-w_0 z_i)}\mathcal{U}_0,$$

or

$$Z \overset{(12.75)}{\simeq} \frac{1}{2z_0}\frac{1}{w_0^{N+1}}\mathcal{U}_0 e^{\mu^2/2-1}\prod_{i\neq0}\frac{1}{(1-w_0 z_i)}, \tag{12.88}$$

where we used $\sqrt{-F_2} = \mu/u_0$ of Eq. (12.75), and we observe that u_0 now drops out. Also note that $F_0 = F(0)$ is evaluated at $u = 0$, *i.e.* at argument $w = z_0^{-1} - u_0 = w_0$ (see Eq. (12.86)). Since μ is very small, we can ignore the factor $e^{\mu^2/2}$. Using Eq. (12.83), *i.e.* $(w_0 z_0)^{-1} = 1 + \bar{n}_0^{-1}$, we have for $N \to \infty$:[‡‡‡]

$$\frac{1}{w_0^{N+1}} = z_0^{N+1}\left(1+\frac{1}{\bar{n}_0}\right)^{N+1} = z_0^{N+1}\left[1+\frac{1}{N}\frac{N}{\bar{n}_0}\right]^{N+1} \simeq z_0^{N+1}e^{N/\bar{n}_0}, \tag{12.89}$$

so that Z becomes

$$Z \propto z_0^N e^{N/\bar{n}_0}e^{-1}\prod_{i\neq0}\frac{1}{(1-w_0 z_i)} = z_0^N e^{(N-\bar{n}_0)/\bar{n}_0}\prod_{i\neq0}\frac{1}{(1-w_0 z_i)}. \tag{12.90}$$

Here $N = \bar{n}_0 + \bar{n}_e$, where the subscript e means "excess" particles (*i.e.* the sum of those in excited states). Then with the approximation $w_0 \simeq z_0^{-1}$, we obtain (for some $i \neq 0$)

$$\bar{n}_{i\neq0} \overset{(12.10)}{=} z_i\frac{d}{dz_i}\ln Z \overset{(12.47)}{=} \frac{w_0 z_i}{1-w_0 z_i} = \frac{1}{z_0/z_i - 1}$$

$$= \frac{1}{e^{(\epsilon_i-\epsilon_0)/kT}-1}, \tag{12.91}$$

and correspondingly (the last line defining $\bar{n}_e$)

$$\bar{n}_0 \overset{(12.90)}{=} z_0\frac{d}{dz_0}\ln Z = N - z_0\frac{d}{dz_0}\sum_{i\neq0}\ln(1-z_i/z_0)$$

$$= N - z_0\sum_{i\neq0}\frac{z_i/z_0^2}{1-z_i/z_0}$$

$$= N - \sum_{i\neq0}\frac{1}{z_0/z_i - 1} \equiv N - \bar{n}_e. \tag{12.92}$$

[‡‡‡] Recall the relation (as given, for instance, in R. Courant [14], Vol. I, p.175)

$$e^{Nz} = \lim_{n\to\infty}\left(1+\frac{Nz}{n}\right)^n.$$

This means, since $\epsilon_{i\neq0} > \epsilon_0$ and therefore

$$\text{with } z_i = e^{-\epsilon_i/kT} \; : \quad \frac{z_0}{z_i} = e^{(\epsilon_i-\epsilon_0)/kT} \to \infty \text{ for } T \to 0,$$

with the temperature approaching absolute zero the particles condense into the ground state with the excess number tending to zero, *i.e.* $\bar{n}_e \to 0$. The fluctuations given by Dingle [17] (p.271) follow from Eq. (12.144) in Example 12.7 together with Eq. (12.92) for (here) $g_j = 1$, *i.e.*

$$i \neq 0 : \quad z_i \frac{d}{dz_i}\bar{n}_i = \bar{n}_i(\bar{n}_i + 1), \tag{12.93a}$$

$$z_0 \frac{d}{dz_0}\bar{n}_0 = \sum_{i\neq0} \bar{n}_i(\bar{n}_i + 1). \tag{12.93b}$$

Concerning Bose–Einstein condensation for an open system, we refer to comments of Dingle [17] (footnote p.268) and Landsberg [38]. In this introductory text to the subject we kept mainly the perfect gas in mind. But we mention that Bose–Einstein condensation has recently also been observed in ^{87}Rb atoms confined to a magnetic trap.[§§§]

12.6 Applications and Examples

Example 12.1: Degeneracy of the s-dimensional harmonic oscillator
The energy levels of the isotropic s-dimensional harmonic oscillator are given by[¶¶¶]

$$E = h\nu\left(N + \frac{1}{2}s\right), \tag{12.94}$$

where s is the number of dimensions, and $N = n_1+n_2+\cdots+n_s$, the n's being integers. Introducing a factor w for each unit contribution to each n, show that the degeneracy of each such energy level is the coefficient of w^N in $(1 + w + w^2 + \cdots)^s$, and hence that it is equal to

$$\frac{(E/h\nu + s/2 - 1)!}{(E/h\nu - s/2)!(s - 1)!}. \tag{12.95}$$

Solution: Although the s-dimensional harmonic oscillator is not an example of the canonical ensemble, we can still use the considerations of Bose–Einstein statistics of Sec. 12.4.2 in an analogous sense. We identify the energy E of Eq. (12.3) with the energy of the harmonic oscillator and thus set

$$E = \sum_i n_i^* \epsilon_i \equiv h\nu\left(n_1 + n_2 + \cdots + n_s + \frac{1}{2}s\right), \quad n_i^* = n_i + \frac{1}{2}. \tag{12.96}$$

[§§§]M.H. Anderson, J.R. Ensher, M.R. Matthews, C.E. Wieman and E.A. Cornell [3]. See also D. Kleppner [37].
[¶¶¶]See *e.g.* A. Messiah [43], Vol. I, Sec. 12.3.1.

Then we define

$$E_N := E - \frac{s}{2}h\nu = h\nu \sum_{i=1}^{s} n_i = Nh\nu, \tag{12.97}$$

so that we have

$$\epsilon_i = h\nu, \quad i = 1, 2, \ldots, s. \tag{12.98}$$

Formally we set

$$z_i = e^{-\epsilon_i/kT} \equiv e^{-h\nu/kT} \equiv z_0, \tag{12.99}$$

and Z_ω of Eq. (12.26) becomes

$$Z_\omega = \prod_{i=1}^{s} \left\{ \sum_{n_i=0}^{\infty} (\omega z_0)^{n_i} \right\} = (1 + \omega z_0 + (\omega z_0)^2 + \cdots)^s, \tag{12.100}$$

with power s since we consider one energy E_N but s different n_i with $n_i = 0, 1, 2, \ldots$. The coefficient of $(\omega z_0)^N$ in Z_ω is the following coefficient which will be derived below:

$$g_N(s) = \frac{(N+s-1)!}{N!(s-1)!} = \frac{(E/h\nu + s/2 - 1)!}{(E/h\nu - s/2)!(s-1)!}. \tag{12.101}$$

This coefficient can be obtained by the considerations leading to Eq. (6.45b). Alternatively one can derive the expression by induction. We can first check the result in simple cases. Thus take the case of $s = 2$ in which

$$\frac{E_N}{h\nu} = n_1 + n_2 = N, \tag{12.102}$$

and we obtain with $g_N(s)$ representing the number of possibilities

$$E_0/h\nu = 0 + 0 \qquad \text{with } g_0(2) = 1,$$
$$E_1/h\nu = 1 + 0, 0 + 1 \qquad \text{with } g_1(2) = 2,$$
$$E_2/h\nu = 2 + 0, 1 + 1, 0 + 2 \text{ with } g_2(2) = 3.$$

From formula (12.101) we obtain

$$g_0(2) = 1, \quad g_1(2) = 2, \quad g_2(2) = 3.$$

For another check see Eq. (6.7) for $N \equiv n, s = 3$.

Next we derive Eq. (12.101). The result $g_N(s)$ is the coefficient of z^N in

$$f_s(z) = (1 + z + z^2 + \cdots)^s.$$

From the theory of functions of a complex variable we recall the *Cauchy formula*[****]

$$\frac{1}{2\pi i} \oint dz \frac{f_s(z)}{z^{N+1}} = \frac{1}{N!} f_s^{(N)}(0). \tag{12.103}$$

Here $f_s^{(N)}(0)$ is the N-th derivative of $f_s(z)$ evaluated at the origin $z = 0$. The term on the right hand side is the coefficient of the $(N+1)$-th term of the Taylor expansion of $f_s(z)$ around $z = 0$. Thus we have

$$g_N(s) = \frac{1}{N!} f_s^{(N)}(0), \tag{12.104}$$

[****]E.G. Phillips [56], p.95.

and it is necessary to compute $f_s^{(N)}(0)$. We have:

$$f_s^{(1)}(z) = sf_{s-1}(z)\left(\sum_{j=1}^{\infty} jz^{j-1}\right), \qquad f_s^{(1)}(0) = sf_{s-1}(0) = s,$$

$$f_s^{(2)}(z) = s(s-1)f_{s-2}(z)\left(\sum_{j=0}^{\infty}(j+1)z^j\right)\left(\sum_{j=0}^{\infty}(j+1)z^j\right)$$

$$+ sf_{s-1}(z)\left(\sum_{j=0}^{\infty}(j+1)(j+2)z^j\right), \quad f_s^{(2)}(0) = s(s+1),$$

$$f_s^{(3)}(z) = \cdots\cdots, \qquad f_s^{(3)}(0) = s(s+1)(s+2). \tag{12.105}$$

Hence

$$f_s^{(N)}(0) = s(s+1)(s+2)\cdots(s+N-1) = \frac{(s+N-1)!}{(s-1)!}, \tag{12.106}$$

and therefore Eq. (12.104) implies[††††]

$$g_N(s) = \frac{(s+N-1)!}{N!(s-1)!}. \tag{12.107}$$

Example 12.2: Asymptotics of degeneracy of s-dimensional oscillator

The degeneracy of the s-dimensional harmonic oscillator is given by the result (12.101). Determine its behaviour for $E/h\nu \gg s$, and for $s \gg E/h\nu$.

Solution: For the case $E/h\nu \gg s$ we use the following relation obtainable with Stirling's approximation of the factorial functions in both numerator and denominator:

$$\frac{i!}{(i+m)!} \overset{i \text{ large}}{\simeq} \frac{1}{i^m}\left[1 - O\left(\frac{1}{i}\right)\right]. \tag{12.108}$$

Applying this relation to $g_N(s)$ of Eq. (12.101) for $E/h\nu$ large, i.e. $E/h\nu \gg s$, we obtain

$$g_N(s) = \frac{(E/h\nu + s/2 - 1)!}{(E/h\nu - s/2)!(s-1)!} = \frac{(E/h\nu + s/2 - 1)!}{(E/h\nu)!}\frac{(E/h\nu)!}{(E/h\nu - s/2)!}\frac{1}{(s-1)!}$$

$$\simeq \frac{(E/h\nu)^{s/2}(E/h\nu)^{s/2}}{(s-1)!} \simeq \frac{(E/h\nu)^s}{(s-1)!}. \tag{12.109}$$

In the case of $s \gg E/h\nu$, we first rewrite the quantity $(E/h\nu - s/2)!$ with the help of the *inversion* or *reflection formula of factorials*,[‡‡‡‡] i.e.

$$(-z)!(z-1)! = \frac{\pi}{\sin \pi z}. \tag{12.110}$$

Thus with $z = s/2 - E/h\nu$, we have:

$$(E/h\nu - s/2)! = \frac{\pi}{(s/2 - E/h\nu - 1)!\sin \pi(s/2 - E/h\nu)}. \tag{12.111}$$

[††††]This is in agreement with literature, see *e.g.* A. Messiah [43], Vol. I, Sec. 12.3.1. Note that $g_N(3) = (N+2)(N+1)/2$ as in Eq. (6.7).

[‡‡‡‡]See *e.g.* W. Magnus and F. Oberhettinger [40], p.1.

Here:

$$\sin \pi(s/2 - E/h\nu) = \sin(\pi s/2)\cos(\pi E/h\nu) - \cos(\pi s/2)\sin(\pi E/h\nu)$$
$$= \begin{cases} s \text{ even}: & \pm \sin(\pi E/h\nu), \\ s \text{ odd}: & \pm \cos(\pi E/h\nu). \end{cases} \tag{12.112}$$

Hence

$$g_N(s) = \frac{(E/h\nu + s/2 - 1)!(s/2 - E/h\nu - 1)!}{\pi(s-1)!} \sin \pi(s/2 - E/h\nu). \tag{12.113}$$

Using the *duplication formula of factorials*,[§§§§]

$$\sqrt{\pi}(2z)! = 2^{2z} z!(z - 1/2)!, \tag{12.114}$$

we can re-express $(s-1)!$ as

$$(s-1)! = \frac{2^{s-1}[(s-1)/2]!(s/2-1)!}{\sqrt{\pi}}. \tag{12.115}$$

With this result, the degeneracy $g_N(s)$ becomes in a good approximation, again using Eq. (12.108),

$$g_N(s) = \frac{(E/h\nu + s/2 - 1)!(s/2 - E/h\nu - 1)!}{\pi 2^{s-1}[(s-1)/2]!(s/2-1)!} \sin \pi(s/2 - E/h\nu)\sqrt{\pi}$$
$$\simeq \frac{(2/s)^{1/2}}{\sqrt{\pi} 2^{s-1}} \sin \pi \left(\frac{s}{2} - \frac{E}{h\nu}\right) \quad \text{for} \quad s \gg \frac{E}{h\nu}. \tag{12.116}$$

This result is practically independent of the energy. Thus the huge numbers involved in the quantum statistical theory of matter quash the large energy dependence of the classical expression (*cf.* Eq. (3.16)).

Example 12.3: Bose–Einstein number of arrangements obtained with ω

Inserting a factor ω for each element introduced, show that the number of distinct arrangements of n indistinguishable elements amongst g degenerate states is equal to the coefficients of ω^n in $(1 + \omega + \omega^2 + \cdots)^g = (1 - \omega)^{-g}$ if any number of elements can be accommodated in any state — Bose–Einstein statistics — and hence that

$$W_{BE} = \prod_i \frac{(g_i + n_i - 1)!}{(g_i - 1)!n_i!}. \tag{12.117}$$

Solution: We obtain from Eq. (12.26):

$$Z_\omega = \prod_i \left(\sum_{n_i}(\omega z_i)^{n_i}\right)^{g_i}$$
$$= \left(1 + \omega z_1 + (\omega z_1)^2 + \cdots\right)^{g_1}\left(1 + \omega z_2 + (\omega z_2)^2 + \cdots\right)^{g_2} \cdots. \tag{12.118}$$

The coefficient of $(wz_1)^{n_1}$ in

$$\left(1 + \omega z_1 + (\omega z_1)^2 + \cdots\right)^{g_1}$$

is (*cf.* Eq. (6.45b) and Eq. (12.101) of Example 12.1)

$$\frac{(g_1 + n_1 - 1)!}{(g_1 - 1)!n_1!}. \tag{12.119}$$

[§§§§]See *e.g.* W. Magnus and F. Oberhettinger [40], p.1.

It follows that the number of ways of arranging n_i elements among g_i states with $N = \sum_i n_i$ for $i = 1, 2, 3, \ldots$, and if any number of indistinguishable elements can be accommodated in any state, the Bose–Einstein distribution is (in agreement with Eq. (6.46))

$$W_{BE} = \prod_i \frac{(g_i + n_i - 1)!}{(g_i - 1)!n_i!}.$$ (12.120)

Example 12.4: Fermi–Dirac number of arrangements obtained with ω

Show that the number of distinct arrangements of n indistinguishable elements amongst g degenerate states is equal to the coefficient of ω^n in $(1+\omega)^g$ if the maximum number of elements which can be accommodated in any state is unity — Fermi–Dirac statistics — and hence that

$$W_{FD} = \prod_i \frac{g_i!}{(g_i - n_i)!n_i!}.$$ (12.121)

Solution: Following the procedure of Example 12.1 with s there replaced by g here, we have

$$
\begin{aligned}
f_g(\omega) &= (1+\omega)^g, \quad f_g(0) = 1, \\
f_g^{(1)}(\omega) &= g f_{g-1}(\omega), \quad f_g^{(1)}(0) = g, \\
f_g^{(2)}(\omega) &= g f_{g-1}^{(1)}(\omega) = g(g-1)f_{g-2}(\omega), \quad f_g^{(2)}(0) = g(g-1), \\
f_g^{(3)}(\omega) &= g(g-1)f_{g-2}^{(1)}(\omega) = g(g-1)(g-2)f_{g-3}(\omega), \\
f_g^{(3)}(0) &= g(g-1)(g-2), \\
f_g^{(n)}(\omega) &= \cdots, \quad f_g^{(n)}(0) = g(g-1)(g-2)\cdots(g-n+1) = \frac{g!}{(g-n)!}.
\end{aligned}
$$ (12.122)

It follows that the coefficient of ω^n in $f_g(\omega)$ is

$$h_n(g) = \frac{1}{n!}f_g^{(n)}(0) = \frac{g!}{(g-n)!n!}.$$ (12.123)

The Fermi–Dirac distribution function is then the product over all such factors, *i.e.*

$$W_{FD} = \prod_i \frac{g_i!}{(g_i - n_i)!n_i!}.$$ (12.124)

This result agrees with that of Eq. (6.44).

Example 12.5: Mean square deviation

Show that

$$\overline{n_j^2} = \frac{1}{Z}\left(z_j \frac{\partial}{\partial z_j}\right)\left(z_j \frac{\partial}{\partial z_j}\right)Z,$$ (12.125)

and hence that the fluctuations in n_j are given by the mean square deviation, also called *variance* $\sigma_{n_j}^2$:

$$\sigma_{n_j}^2 := \overline{(n_j - \overline{n_j})^2} = \overline{n_j^2} - (\overline{n_j})^2 = z_j \frac{\partial}{\partial z_j}\overline{n_j}.$$ (12.126)

The quantity σ_{n_j} is called *standard deviation*, and the overline denotes the average over the distribution.

Solution: Using the definition (12.8) of a mean, we have ($\sum$ meaning $\sum_i n_i = N$)

$$Z\overline{n_j^2} = \sum n_j^2 \prod_i z_i^{n_i} = \sum n_j \left(n_j \prod_i z_i^{n_i} \right) = \sum n_j \left(n_j z_j^{n_j} \prod_{i \neq j} z_i^{n_i} \right)$$

$$= \sum n_j z_j \frac{\partial}{\partial z_j} \prod_i z_i^{n_i} = \sum z_j \frac{\partial}{\partial z_j} n_j \prod_i z_i^{n_i}$$

$$= z_j \frac{\partial}{\partial z_j} \left\{ \sum n_j \prod_i z_i^{n_i} \right\} \overset{(12.9)}{=} \left(z_j \frac{\partial}{\partial z_j} \right) \left(z_j \frac{\partial}{\partial z_j} Z \right). \tag{12.127}$$

Hence

$$\overline{n_j^2} = \frac{1}{Z} \left(z_j \frac{\partial}{\partial z_j} \right) \left(z_j \frac{\partial}{\partial z_j} \right) Z. \tag{12.128}$$

It follows also that (using Eq. (12.10) for $\overline{n}_j$)

$$\overline{n_j^2} - \overline{n}_j^2 = \frac{1}{Z} \left(z_j \frac{\partial}{\partial z_j} \right) \left(z_j \frac{\partial}{\partial z_j} \right) Z - \left(\frac{1}{Z} z_j \frac{\partial}{\partial z_j} Z \right) \left(\frac{1}{Z} z_j \frac{\partial}{\partial z_j} Z \right). \tag{12.129}$$

In order to prove the next relation we begin with

$$Z(n_j - \overline{n_j}) = n_j \sum \prod_i z_i^{n_i} - \sum n_j \prod_i z_i^{n_i} = n_j Z - \left(z_j \frac{\partial}{\partial z_j} Z \right), \tag{12.130}$$

from which we obtain

$$Z^2 (n_j - \overline{n_j})^2 = \left(z_j \frac{\partial}{\partial z_j} Z \right)^2 - 2 n_j Z \left(z_j \frac{\partial}{\partial z_j} Z \right) + (n_j Z)^2. \tag{12.131}$$

From this relation we obtain by taking the mean value:

$$\overline{Z^2 (n_j - \overline{n_j})^2} = \left(z_j \frac{\partial}{\partial z_j} Z \right)^2 + \overline{(n_j Z)^2} - \overline{2 n_j Z \left(z_j \frac{\partial}{\partial z_j} Z \right)}$$

$$= \left(z_j \frac{\partial}{\partial z_j} Z \right)^2 + Z^2 \overline{n_j^2} - 2 \overline{n_j} Z \left(z_j \frac{\partial}{\partial z_j} Z \right)$$

$$= \left(z_j \frac{\partial}{\partial z_j} Z \right)^2 + Z^2 \overline{n_j^2} - 2(\overline{n_j} Z)^2 = Z^2 (\overline{n_j^2} - \overline{n_j}^2). \tag{12.132}$$

The remaining part of the problem is a matter of differentiation. We have:

$$z_j \frac{\partial}{\partial z_j} \overline{n_j} = z_j \frac{\partial}{\partial z_j} \left[\frac{1}{Z} z_j \left(\frac{\partial Z}{\partial z_j} \right) \right]$$

$$= z_j \left[- \left(\frac{1}{Z^2} \frac{\partial Z}{\partial z_j} \right) z_j \left(\frac{\partial Z}{\partial z_j} \right) + \frac{1}{Z} \left(\frac{\partial Z}{\partial z_j} \right) + \frac{1}{Z} z_j \frac{\partial^2 Z}{\partial z_j^2} \right]. \tag{12.133}$$

On the other hand:

$$\overline{n_j^2} - \overline{n_j}^2 = \frac{1}{Z} \left(z_j \frac{\partial}{\partial z_j} \right) \left(z_j \frac{\partial}{\partial z_j} \right) Z - \left(\frac{1}{Z} z_j \frac{\partial}{\partial z_j} Z \right)^2$$

$$= \frac{1}{Z} z_j \left[\frac{\partial}{\partial z_j} \left(z_j \frac{\partial Z}{\partial z_j} \right) \right] - \left(\frac{1}{Z} z_j \frac{\partial Z}{\partial z_j} \right)^2$$

$$= \frac{z_j}{Z} \left[\left(\frac{\partial Z}{\partial z_j} \right) + z_j \frac{\partial^2 Z}{\partial z_j^2} \right] - \left(\frac{z_j}{Z} \frac{\partial Z}{\partial z_j} \right)^2 = z_j \frac{\partial}{\partial z_j} \overline{n_j}. \tag{12.134}$$

This expression is seen to be identical with that above, which had to be shown.

Example 12.6: Correlation between occupation numbers

Show that the correlations between n_i and n_j are given by

$$M := \overline{(n_i - \overline{n_i})(n_j - \overline{n_j})} = z_i \frac{\partial}{\partial z_i} \overline{n_j} = z_j \frac{\partial}{\partial z_j} \overline{n_i}. \tag{12.135}$$

Solution: We have as in Eq. (12.130)

$$n_i - \overline{n_i} = n_i - \frac{1}{Z}\left(z_i \frac{\partial Z}{\partial z_i}\right). \tag{12.136}$$

Hence

$$M \equiv \overline{(n_i - \overline{n_i})(n_j - \overline{n_j})} = \overline{\left(n_i - \frac{1}{Z} z_i \frac{\partial Z}{\partial z_i}\right)\left(n_j - \frac{1}{Z} z_j \frac{\partial Z}{\partial z_j}\right)}$$

$$= \overline{n_i n_j} - \frac{\overline{n_i}}{Z} z_j \frac{\partial Z}{\partial z_j} - \frac{\overline{n_j}}{Z} z_i \frac{\partial Z}{\partial z_i} + \frac{z_i z_j}{Z^2}\frac{\partial Z}{\partial z_i}\frac{\partial Z}{\partial z_j}. \tag{12.137}$$

Here as in Eq. (12.8)

$$Z \overline{n_i n_j} = \sum n_i n_j \prod_k z_k^{n_k} = \sum n_i \left(n_j \prod_k z_k^{n_k}\right) = \sum n_i z_j \frac{\partial}{\partial z_j}\left(\prod_k z_k^{n_k}\right)$$

$$= z_j \frac{\partial}{\partial z_j}\left\{\left(\sum n_i \prod_k z_k^{n_k}\right)\right\} = \left(z_j \frac{\partial}{\partial z_j}\right)\left(z_i \frac{\partial}{\partial z_i}\right) Z, \tag{12.138}$$

$$\overline{n_i n_j} = \frac{1}{Z}\left(z_j \frac{\partial}{\partial z_j}\right)\left(z_i \frac{\partial}{\partial z_i}\right) Z. \tag{12.139}$$

Considering the next term in M we have

$$\frac{\overline{n_i}}{Z} z_j \frac{\partial Z}{\partial z_j} = \frac{1}{Z}\overline{n_i z_j \frac{\partial Z}{\partial z_j}} \overset{(12.8)}{=} \frac{1}{Z^2}\sum n_i z_j \frac{\partial Z}{\partial z_j}\prod_k z_k^{n_k}$$

$$= \frac{1}{Z^2} z_j \frac{\partial Z}{\partial z_j}\sum n_i \prod_k z_k^{n_k} \overset{(12.10)}{=} \frac{1}{Z^2}\left(z_j \frac{\partial Z}{\partial z_j}\right)\left(z_i \frac{\partial Z}{\partial z_i}\right). \tag{12.140}$$

It follows that

$$M = \frac{1}{Z}\left(z_j \frac{\partial}{\partial z_j}\right)\left(z_i \frac{\partial}{\partial z_i}\right) Z - \frac{2}{Z^2}\left(z_j \frac{\partial Z}{\partial z_j}\right)\left(z_i \frac{\partial Z}{\partial z_i}\right) + \frac{1}{Z^2} z_i z_j \left(\frac{\partial Z}{\partial z_i}\right)\left(\frac{\partial Z}{\partial z_j}\right)$$

$$= \frac{1}{Z}\left(z_j \frac{\partial}{\partial z_j}\right)\left(z_i \frac{\partial}{\partial z_i}\right) Z - \frac{1}{Z^2}\left(z_j \frac{\partial Z}{\partial z_j}\right)\left(z_i \frac{\partial Z}{\partial z_i}\right)$$

$$= \frac{1}{Z}\left(z_j \frac{\partial z_i}{\partial z_j}\right)\left(\frac{\partial Z}{\partial z_i}\right) + \frac{1}{Z}\left(z_j z_i \frac{\partial^2 Z}{\partial z_j \partial z_i}\right) - \frac{1}{Z^2}\left(z_j \frac{\partial Z}{\partial z_j}\right)\left(z_i \frac{\partial Z}{\partial z_i}\right)$$

$$= \frac{z_i}{Z}\frac{\partial Z}{\partial z_i} + \frac{1}{Z} z_j z_i \frac{\partial^2 Z}{\partial z_j \partial z_i} - \frac{1}{Z^2}\left(z_j \frac{\partial Z}{\partial z_j}\right)\left(z_i \frac{\partial Z}{\partial z_i}\right)$$

$$= \frac{z_i}{Z}\frac{\partial Z}{\partial z_i} + z_i z_j \frac{\partial}{\partial z_i}\left(\frac{1}{Z}\frac{\partial Z}{\partial z_j}\right) = z_i \frac{\partial}{\partial z_i}\left(\frac{z_j}{Z}\frac{\partial Z}{\partial z_j}\right) = z_i \frac{\partial}{\partial z_i}\overline{n_j}. \tag{12.141}$$

Example 12.7: Fluctuation for a system in a heat bath

Calculate the fluctuation in n_j for a *classical system* in a heat bath, *i.e.* for *Maxwell–Boltzmann statistics*. What are the corresponding fluctuations in *Bose–Einstein* and *Fermi–Dirac statistics*?

Solution: We begin with the fluctuation in n_j for a classical system in a heat bath, and we recall that $z_i = e^{-\epsilon_i/kT}$. From Eq. (12.126) we know that the fluctuation is given by the following expression which we have to evaluate for the present case:

$$
\overline{(n_j - \overline{n_j})^2} = z_j \frac{\partial}{\partial z_j} \overline{n_j} \overset{(12.19)}{=} z_j \frac{\partial}{\partial z_j} \left(N \frac{g_j z_j}{\sum_i g_i z_i} \right) = z_j N \frac{g_j}{\sum_i g_i z_i} - N z_j g_j z_j \frac{g_j}{(\sum_i g_i z_i)^2}
$$

$$
= N \frac{g_j z_j}{\sum_i g_i z_i} - N \frac{g_j^2 z_j^2}{(\sum_i g_i z_i)^2} = \overline{n_j} - \frac{1}{N}(\overline{n_j})^2. \tag{12.142}
$$

In the case of *Bose–Einstein statistics* we have:

$$
\overline{n_j} \overset{(12.48)}{=} \frac{g_j}{w_0^{-1} z_j^{-1} - 1} = \frac{g_j w_0 z_j}{1 - w_0 z_j} \qquad \text{and} \qquad \overline{n_j} + g_j = \frac{g_j w_0 z_j}{1 - w_0 z_j} + g_j = \frac{g_j}{1 - w_0 z_j}, \tag{12.143}
$$

so that

$$
\overline{(n_j - \overline{n_j})^2} = z_j \frac{\partial}{\partial z_j} \overline{n_j} = z_j \frac{\partial}{\partial z_j} \left[\frac{g_j w_0 z_j}{1 - w_0 z_j} \right] = \frac{z_j g_j w_0 (1 - w_0 z_j) + z_j g_j w_0^2 z_j}{(1 - w_0 z_j)^2}
$$

$$
= \frac{z_j g_j w_0}{(1 - w_0 z_j)^2} = \frac{\overline{n_j}}{1 - w_0 z_j} = \frac{1}{g_j} \overline{n_j} (\overline{n_j} + g_j). \tag{12.144}
$$

In the case of *Fermi–Dirac statistics* we have correspondingly:

$$
\overline{n_j} \overset{(12.48)}{=} \frac{g_j w_0 z_j}{1 + w_0 z_j} \qquad \text{and} \qquad \overline{n_j} - g_j = \frac{g_j w_0 z_j - g_j (1 + w_0 z_j)}{1 + w_0 z_j} = -\frac{g_j}{1 + w_0 z_j}, \tag{12.145}
$$

so that

$$
\overline{(n_j - \overline{n_j})^2} = z_j \frac{\partial}{\partial z_j} \left[\frac{g_j w_0 z_j}{1 + w_0 z_j} \right] = \frac{z_j g_j w_0 (1 + w_0 z_j) - g_j w_0^2 z_j^2}{(1 + w_0 z_j)^2}
$$

$$
= \frac{\overline{n_j}}{1 + w_0 z_j} = \overline{n_j} \left(\frac{\overline{n_j} - g_j}{-g_j} \right) = \frac{1}{g_j} \overline{n_j} (g_j - \overline{n_j}) > 0. \tag{12.146}
$$

Here $g_j \geq 1$ and $0 \leq \overline{n_j} \leq 1$.

Example 12.8: Stirling's formula by the method of steepest descents

Use the method of steepest descents to evaluate

$$
n! = \int_0^\infty x^n e^{-x} dx, \tag{12.147}
$$

and hence obtain *Stirling's formula*

$$
n! \approx \sqrt{2\pi} n^{n+1/2} e^{-n}. \tag{12.148}
$$

Solution: The method of steepest descents evaluates the contour integral

$$\oint_C e^{f(z)} dz \qquad (12.149)$$

by expanding $f(z)$ about its extremum at z_0 with $z - z_0 = ix$, $dz = idx$ (observe that this implies an integration parallel to the axis of imaginary z through the saddle point). Then, since $f'(z_0) = 0$, and $(z - z_0)^2 = -x^2$,

$$\oint e^{f(z)} dz \simeq e^{f(z_0)} i \int_{-\infty}^{\infty} e^{-x^2 f''(z_0)/2} dx \stackrel{(2.20)}{=} i\sqrt{\frac{2\pi}{f''(z_0)}} e^{f(z_0)}. \qquad (12.150)$$

The gamma function $\Gamma(n+1)$ or factorial function $n!$ is defined as

$$n! = \int_0^{\infty} e^{-z} z^n dz = \int_0^{\infty} e^{-z + n \ln z} dz. \qquad (12.151)$$

Thus here

$$f(z) = -z + n \ln z, \quad z_0 = n, \quad f''(z_0) = -\frac{n}{z_0^2} = -\frac{1}{n}, \qquad (12.152)$$

and hence integrating as described above through the saddle point, one obtains

$$n! \simeq \sqrt{2\pi} n^{n+1/2} e^{-n}, \qquad (12.153)$$

where we used Eq. (2.20) (with doubled range of integration),

$$\int_{-\infty}^{\infty} dx e^{-\omega^2 x^2/2} = \sqrt{\frac{2\pi}{\omega^2}}.$$

Example 12.9: Coefficient of ω^N in Z_ω^{MB} by steepest descent
Find by the method of steepest descents the coefficient of ω^N in

$$Z_\omega^{MB} = N! e^{\omega \sum_i g_i z_i}, \qquad (12.154)$$

and compare the answer with the exact result.

Solution: As described in the text and given by Eq. (12.32), the coefficient of ω^N in Z_ω^{MB} is given by

$$\frac{1}{2\pi i} \oint \frac{1}{\omega^{N+1}} Z_\omega^{MB} d\omega = \frac{N!}{2\pi i} \oint \frac{e^{\omega \sum_i g_i z_i}}{\omega^{N+1}} d\omega \equiv \frac{N!}{2\pi i} \oint e^{f(\omega)} d\omega. \qquad (12.155)$$

Here

$$f(\omega) = \omega \sum_i g_i z_i - (N+1) \ln \omega,$$

$$f'(\omega) = \sum_i g_i z_i - \frac{N+1}{\omega} \quad \text{implying} \quad f'(\omega_0) = 0 \quad \text{with} \quad \omega_0 = \frac{N+1}{\sum_i g_i z_i},$$

$$f''(\omega) = \frac{N+1}{\omega^2}. \qquad (12.156)$$

Using now Eq. (12.150) of Example 12.8 on the method of steepest descents, and from above that $f(w_0) = (N+1) - (N+1)\ln w_0$, the coefficient of w^N in Z_w^{MB} is (in the approximation of that method) given by

$$
\simeq \frac{N!}{2\pi i}i\sqrt{\frac{2\pi}{(N+1)/w_0^2}}e^{(N+1)-(N+1)\ln w_0}
$$

$$
= \frac{N!}{\sqrt{2\pi}}\frac{w_0}{\sqrt{N+1}}e^{(N+1)}w_0^{-(N+1)} = \frac{N!}{\sqrt{2\pi(N+1)}}e^{(N+1)}w_0^{-N}
$$

$$
= \frac{N!}{\sqrt{2\pi(N+1)}}e^{(N+1)}\left(\frac{\sum_i g_i z_i}{N+1}\right)^N
$$

$$
= \frac{N!}{\sqrt{2\pi(N+1)}}(N+1)^{-N}e^{(N+1)}\left(\sum_i g_i z_i\right)^N
$$

$$
\underset{\text{Stirling approximation}}{\simeq} \left(\sum_i g_i z_i\right)^N. \qquad (12.157)
$$

Thus, although we used the approximation of the method of steepest descent on the way, and with this the Stirling formula, the final result is seen to be identical with the exact result given by Eq. (12.18).

Example 12.10: Estimate of correction term
Find expressions, in the various statistics, for the first correction term $-(1/2)\ln(2\pi f''(w_0))$ to $\ln Z$ in terms of the energy levels and their degeneracies. Explain why this term is *normally* negligible compared with the leading term in $\ln Z$.

Solution: We consider *Maxwell–Boltzmann statistics*. In this case we obtain from Example 12.9 the following equations:

$$
f(w_0) = w_0 \sum_i g_i z_i - (N+1)\ln w_0,
$$

$$
w_0 = \frac{N+1}{\sum_i g_i z_i}, \quad \text{implying} \quad w_0 \sum_i g_i z_i = N+1,
$$

$$
f''(w_0) = \frac{N+1}{w_0^2}. \qquad (12.158)
$$

Therefore we obtain for $-(1/2)\ln[2\pi f''(w_0)]$ from Eq. (12.43):

$$
-\frac{1}{2}\ln[2\pi f''(w_0)] = \ln Z - f(w_0) \qquad (12.159)
$$

with

$$
L.H.S. = -\frac{1}{2}\ln\left[2\pi\frac{N+1}{w_0^2}\right] = -\frac{1}{2}\ln[2\pi(N+1)] + \ln w_0, \qquad (12.160)
$$

and with Eq. (12.18):

$$
R.H.S. = \ln Z - f(w_0) = N\ln\left(\sum_i g_i z_i\right) - f(w_0)
$$

$$
= N\ln\left(\sum_i g_i z_i\right) - [(N+1) - (N+1)\ln w_0]. \qquad (12.161)
$$

Equating both sides we obtain:

$$-\frac{1}{2}\ln[2\pi(N+1)] + \ln\omega_0 = N\ln\left(\sum_i g_i z_i\right) - (N+1) + (N+1)\ln\omega_0,$$

and hence (for N large)

$$-\frac{1}{2}\ln[2\pi(N+1)] = N\ln\left(\sum_i g_i z_i\right) - (N+1) + N\ln\omega_0,$$

$$-N\ln\omega_0 = \ln\left(\sum_i g_i z_i\right)^N + \ln\sqrt{2\pi(N+1)} - (N+1)$$

$$= \ln\left[\sqrt{2\pi(N+1)}\left(\sum_i g_i z_i\right)^N\right] - (N+1),$$

$$N\ln\omega_0 = (N+1) - \ln\left[\sqrt{2\pi(N+1)}\left(\sum_i g_i z_i\right)^N\right],$$

$$\ln\omega_0 = \frac{N+1}{N} - \frac{1}{N}\ln\left[\sqrt{2\pi(N+1)}\left(\sum_i g_i z_i\right)^N\right]. \tag{12.162}$$

From this relation we obtain ω_0, *i.e.*

$$\omega_0 = e^{(N+1)/N}\{2\pi(N+1)\}^{-1/2N}\left(\sum_i g_i z_i\right)^{-1},$$

$$\frac{1}{\omega_0} = \left(\sum_i g_i z_i\right)\{2\pi(N+1)\}^{1/2N} e^{-(N+1)/N}, \tag{12.163}$$

and hence, by squaring and multiplying by $2\pi(N+1)$,

$$\underbrace{2\pi\frac{N+1}{\omega_0^2}}_{2\pi f''(\omega_0)} = \left(\sum_i g_i z_i\right)^2\{2\pi(N+1)\}^{(N+1)/N} e^{-2(N+1)/N}$$

$$= \left(\sum_i g_i z_i\right)^2\{\sqrt{2\pi(N+1)}e^{-1}\}^{2(N+1)/N}. \tag{12.164}$$

Thus

$$-\frac{1}{2}\ln[2\pi f''(\omega_0)] = -\ln\left(\sum_i g_i z_i\right) - \frac{N+1}{N}\ln\{\sqrt{2\pi(N+1)}e^{-1}\}. \tag{12.165}$$

Since $z_i = e^{-\epsilon_i/kT}$, this result expresses the correction term in terms of energies ϵ_i and degeneracies g_i. We know from Eq. (6.58) or Example 3.5 that $\sum_i g_i z_i \propto V$. Therefore

$$Z \overset{(12.18)}{=} \left(\sum_i g_i z_i\right)^N \propto V^N, \quad \ln Z \propto N\ln V, \quad \ln\left(\sum_i g_i z_i\right) \propto \ln V. \tag{12.166}$$

Thus, since the leading term in $\ln Z$, *i.e.* $f(\omega_0) \sim \omega_0\sum_i g_i z_i$, is proportional to V, the first correction term to $\ln Z \sim \ln V$ is normally negligible as claimed between Eqs. (12.46) and (12.47) (there in the context of the Bose–Einstein distribution). The calculations in the cases of Fermi–Dirac and Bose–Einstein statistics proceed accordingly.

12.7   Problems without Worked Solutions

Example 12.11: Operator conversion

Show that[¶¶¶¶]

$$\left(z_j \frac{\partial}{\partial z_j}\right)\overline{n}_j = \left(-\frac{\partial}{\partial(\epsilon_j/kT)}\right)\overline{n}_j, \tag{12.167}$$

and correspondingly

$$\overline{n_j^2} - \overline{n}_j{}^2 = \frac{\partial^2 \ln Z}{\partial(\epsilon_j/kT)^2}. \tag{12.168}$$

Example 12.12: Calculation of mean values

Generalizing Eqs. (12.10), (12.125), one has *e.g.*

$$\overline{n_i n_j n_k} = \frac{1}{Z}\left(z_i \frac{\partial}{\partial z_i}\right)\left(z_j \frac{\partial}{\partial z_j}\right)\left(z_k \frac{\partial}{\partial z_k}\right)Z. \tag{12.169}$$

Verify this relation (parallel to Example 12.5) for the case:

$$\overline{n_j^3} = \frac{1}{Z}\left(z_j \frac{\partial}{\partial z_j}\right)\left(z_j \frac{\partial}{\partial z_j}\right)\left(z_j \frac{\partial}{\partial z_j}\right)Z. \tag{12.170}$$

Example 12.13: Fluctuation in Maxwell–Boltzmann statistics

Show that in the case of Maxwell–Boltzmann statistics the relative square of the fluctuation of the occupation number is

$$\frac{\overline{n_j^2} - \overline{n}_j{}^2}{\overline{n}_j{}^2} \simeq \frac{1}{\overline{n}_j}. \tag{12.171}$$

Example 12.14: Fluctuation in the case of photons

Show that in the case of nonconserved particles obeying Bose–Einstein statistics (like photons), the relative square of the particle number fluctuation is

$$\frac{\overline{n_j^2} - \overline{n}_j{}^2}{\overline{n}_j{}^2} \simeq 1 + \frac{1}{\overline{n}_j}. \tag{12.172}$$

What do you conclude from the difference between this result and that of Example 12.13? (For discussion see *e.g.* Reif [59], Sec. 9.5).

Example 12.15: Fluctuation in n_0

Calculate the fluctuation in n_0 by verifying Eq. (12.93b), *i.e.*

$$\sigma_{n_0}^2 = z_0 \frac{\partial}{\partial z_0}\overline{n}_0 = \sum_{i \neq 0}\overline{n}_i(\overline{n}_i + 1), \qquad g_i = 1. \tag{12.173}$$

Example 12.16: Correlations for a system in a heat bath

Calculate the correlation between n_i and $n_j, i \neq j$, for a system in a heat bath for Maxwell–Boltzmann, Bose–Einstein and Fermi–Dirac statistics.

[¶¶¶¶]The expression on the right is consistently used in the text of F. Reif [59]. See *e.g.* Sec. 9.2.

Example 12.17: Critical temperature of ideal Bose–Einstein gas

Starting from the Bose–Einstein distribution for the mean number $\bar{n}_i$ of (spin zero) He^4-molecules in a certain state i, and the number of translational states g_i (or *a priori* probability as considered in Sec. 3.4), consider the expression for the total number of such molecules N in a volume V which as explained after Eq. (12.44) determines the value of $\omega_0 = \exp(\mu/kT)$, or the *chemical potential* μ (*cf.* Eq. (12.49)). Show that one must have $\mu < 0$ and $\partial\mu/\partial T < 0$. At a critical temperature T_c the quantity μ attains its maximum value zero. Obtain the critical temperature at this point. Evaluate the integral involved by expansion of the denominator, *i.e.*[*****]

$$\int_0^\infty \frac{\sqrt{x}\,dx}{e^x - 1} = \int_0^\infty \sqrt{x}(e^{-x} + e^{-2x} + \cdots)dx = \int_0^\infty \sqrt{x}e^{-x}\left(1 + \frac{1}{2^{3/2}} + \frac{1}{3^{3/2}} + \cdots\right)dx$$

$$= 2\int_0^\infty w^2 e^{-w^2}\,dw \sum_{i=1}^\infty \frac{1}{i^{3/2}} = 2.61\frac{\sqrt{\pi}}{2} \simeq 2.31. \tag{12.174}$$

The transition to Bose–Einstein condensation occurs below this critical temperature. [Hint: Consider $dN(\mu, T) = 0$]. (Answer: $T_c = (\sqrt{2}\pi^2/2.31)^{2/3}(\hbar^2/km)(N/V)^{2/3}$).

Example 12.18: Partition function of s-fold harmonic oscillator

In Example 12.1 the degeneracy $g_j(s)$ of the s-fold simple harmonic oscillator was calculated. Consider now the partition function

$$Z_s = \sum_j g_j e^{-E_j/kT}, \quad E_j = h\nu\left(j + \frac{1}{2}s\right), \quad j = 0, 1, 2, \ldots . \tag{12.175}$$

(a) Show that the canonical partition function at temperature T is given by

$$Z_s = \left(\frac{1}{2\sinh x}\right)^s, \quad x = \frac{h\nu}{2kT}. \tag{12.176}$$

What is its mean energy E_s?
(b) Derive the classical partition function and the mean energy of a one-dimensional oscillator of mass m and natural frequency ν. Verify that Z_s, E_s lead to these expressions in the classical limit $T \to \infty$.

Example 12.19: Density of condensed He^4 molecules

Consider an ideal Bose–Einstein gas, specifically of He^4 molecules, in which sufficient condensation has occurred that most of the molecules are in their ground state. Let the gas be contained in a rectangular vessel and let a height z be measured upwards from the bottom of the vessel. Where in the vessel do most of the molecules assemble (a) with no gravitational potential $V = mgz$ acting, and (b) with this? [Hint: Case (b) requires the solution of the Schrödinger equation in terms of Airy functions].[†††††]

Example 12.20: Bose–Einstein condensation

A simple application of the saddle-point method becomes invalid at the Bose–Einstein condensation temperature due to the saddle-point in the evaluation of the partition function

$$\frac{1}{2\pi i}\oint_{(\omega=0)} \frac{d\omega}{\omega^{N+1}\prod_{\text{energy levels}}(1 - \omega z_i)^{g_i}}, \quad z_i = e^{-\epsilon_i/kT},$$

[*****]See Ya.P. Terletskii [75], p.205.

[†††††]See also H.J.G. Hayman [31], pp.153-154. For Airy functions see also O. Vallée and M. Soares [80].

coming to lie very close to the pole at

$$\omega = \frac{1}{z_1} = e^{\epsilon_1/kT},$$

which can be taken numerically as 1 without loss of generality by defining the first energy level as $\epsilon_1 = 0$. Apply the method of Van der Waerden [81] for dealing with a saddle point which approaches a pole. The method is also explained in Temme [74], Chapter 21. (This is more a research problem than an exercise).

Chapter 13

The Boltzmann Transport Equation

13.1   Introductory Remarks

In Chapter 10 we considered electrons in metals. In the present chapter we return to this topic since this is a central application of statistical distribution functions, but we approach this topic from a more general consideration of distribution functions, namely as solutions of the Boltzmann transport equation. We then derive for instance expressions for the electrical conductivity of metals (for the formulation of which textbooks of electrodynamics resort to simplified semi-classical models $i.e.$ without the use of the Fermi–Dirac distribution function), and study the temperature dependence of relaxation times and hence of the resistance of a solid. For a wider introduction to the subject we recommend consultation of the book of Reif [59], Chapters 13 and 14, and the highly readable text of Wannier [83]. Books which deal with the subject on a more advanced level are $e.g.$ those of Cercignani, [11] and of Duderstadt and Martin [22].

13.2   Distribution Functions

Let n (we drop the subscript i for convenience) be the number of given particles in a volume V (with total number of particles N), such as electrons, having a given property like those which have an energy ϵ within a certain range. Then ($cf.$ $e.g.$ Eq. (10.2))

$$n = f \times g, \qquad (13.1)$$

215

where f is the distribution function, *i.e.* the fraction of those states actually occupied by electrons, and g is the statistical weight, *i.e.* the number of wave mechanical states. The number g is the *a priori* probability,

$$g \propto \triangle q \triangle p \quad \text{per direction per electron.} \tag{13.2}$$

As we observed in Chapter 3 in our consideration of the Liouville equation, this quantity g must be a choice independent of time. This is so since we have no information on the external conditions and we do not know the time. Thus for a consistent choice of g, time t cannot enter. In the case of the 3-dimensional system of electrons, g is given as

$$g = \frac{2V\, dp_x dp_y dp_z}{h^3}, \tag{13.3}$$

where V is the total volume. From Eq. (13.1) we obtain[*]

$$f = \frac{n}{g}. \tag{13.4}$$

In our previous consideration no mention was made of electric or other fields. Thus with no such fields present, we have $f \to f_0$, where f_0 is the distribution function of Fermi–Dirac particles,

$$f_0 = \frac{1}{e^{(\epsilon - \epsilon_0)/kT} + 1}. \tag{13.5}$$

Here we deal only with stationary quantum mechanical states, therefore under steady conditions (*e.g.* dynamic equilibrium) the number of electrons n having a given property is independent of time. Since by Liouville's theorem of Chapter 3 g is also independent of time, we conclude that f is independent of time for a steady state system. It follows that $df/dt = 0$, or

$$f(t + dt, \mathbf{v} + \dot{\mathbf{v}}dt, \mathbf{r} + \mathbf{v}dt) = f(t, \mathbf{v}, \mathbf{r}),$$

or

$$\frac{\partial f}{\partial t} + \dot{\mathbf{v}} \cdot \nabla_v f + \mathbf{v} \cdot \nabla_r f = 0, \tag{13.6}$$

where

$$\dot{\mathbf{v}} \cdot \nabla_v f = \dot{v}_x \frac{\partial f}{\partial v_x} + \cdots + \cdots .$$

Equation (13.6) is exact but useless — the latter because it includes the acceleration due to collisions, due to external electric and magnetic fields *etc.*

---

[*]See *e.g.* L.E. Reichl [57], p.469.

It is more convenient to mean by the acceleration $\dot{\mathbf{v}}$ only acceleration which is due to the external fields. Hence we throw out contributions resulting from collisions by writing the equation in the form

$$\frac{\partial f}{\partial t} + \dot{\mathbf{v}} \cdot \nabla_v f + \mathbf{v} \cdot \nabla_r f = \left(\frac{\partial f}{\partial t}\right)_{\text{collisions}}, \tag{13.7}$$

where $\dot{\mathbf{v}}$ now includes only the acceleration due to external fields. This equation is called the *exact Boltzmann transport equation*.

It is clear that $(\partial f/\partial t)_{\text{collisions}}$ is the value of $(\partial f/\partial t)$ when (i) $\dot{\mathbf{v}} = 0$, *i.e.* when there are no external fields, and (ii) when $\nabla_r f = 0$, *i.e.* when the conditions are uniform. Suppose now that these conditions are disturbed. Then what is the rate at which the system relaxes the distribution function f to f_0, the Fermi–Dirac distribution function (13.5)? Very often

$$\left(\frac{\partial f}{\partial t}\right)_{\text{collisions}} \propto -(f - f_0). \tag{13.8}$$

Then it is convenient to write

$$\left(\frac{\partial f}{\partial t}\right)_{\text{collisions}} = -\frac{f - f_0}{\tau}, \tag{13.9}$$

where τ is called the *time of relaxation* or collision time (*i.e.* the mean time in which f sinks back to f_0). On this assumption the Boltzmann transport equation becomes[†]

$$\frac{\partial f}{\partial t} + \dot{\mathbf{v}} \cdot \nabla_v f + \mathbf{v} \cdot \nabla_r f = -\frac{f - f_0}{\tau}. \tag{13.10}$$

In the next section we solve this equation for two typical cases.

13.3   Solution of the Boltzmann Equation

In this section we solve Eq. (13.10) for two typical cases. Having found the solution $f - f_0$ for these, we calculate the current density $\mathbf{j}$ of the electrons and hence the electrical conductivity σ. Thereafter we consider the application of the results to metals.

[†] *Cf.* F. Reif [59], Eq. (13.6.3).

13.3.1 Solving the Boltzmann equation for two typical cases

The two cases are related.

(i) The first case we consider is that of a uniform steady external electric field $\mathbf{F}$ in the x-direction. Here the word uniform implies independence of position so that $\boldsymbol{\nabla}_r \to 0$, and the word steady implies independence of time so that $\partial/\partial t \to 0$. It follows that Eq. (13.10) reduces to[‡]

$$\dot{\mathbf{v}} \cdot \boldsymbol{\nabla}_v f = -\frac{f - f_0}{\tau}, \qquad f - f_0 = -\tau \dot{\mathbf{v}} \cdot \boldsymbol{\nabla}_v f. \tag{13.11}$$

Since the field F acts in the direction of x we have

$$m\dot{v}_x = eF, \tag{13.12}$$

where m is throughout the "*effective mass*" of the electron (often written m^* and discussed in detail in [46], and is here taken as its ordinary mass). Hence we have the solution

$$f - f_0 = -\frac{eF\tau}{m} \frac{\partial f}{\partial v_x}. \tag{13.13}$$

To a first approximation we can replace on the right hand side f by f_0, so that

$$f - f_0 \simeq -\frac{eF\tau}{m} \frac{\partial f_0}{\partial v_x}. \tag{13.14}$$

As a second approximation we obtain from Eq. (13.13):

$$f - f_0 = -\frac{eF\tau}{m} \frac{\partial}{\partial v_x} \left\{ f_0 - \frac{eF\tau}{m} \frac{\partial f_0}{\partial v_x} \right\}. \tag{13.15}$$

The new correction term proportional to F^2 is nearly always negligible. Hence we use the approximation (13.14). Now, f_0 is the ordinary Fermi–Dirac distribution function as applicable to isotropic systems, and since the energy of the electron is proportional to v^2, and f_0 is a function of energy, we have

$$f_0 = f_0(v^2) \qquad \text{only.} \tag{13.16}$$

Since

$$v^2 = v_x^2 + v_y^2 + v_z^2, \qquad v\,dv = v_x\,dv_x,$$

we obtain

$$f - f_0 = -\frac{eF\tau}{m} \frac{v_x}{v} \frac{\partial f_0}{\partial v}. \tag{13.17}$$

[‡] *Cf.* G.H. Wannier [83], p.436.

(ii) The second case we consider is that of a uniform oscillatory external electric field F,

$$F \propto e^{i\omega t}, \qquad \therefore \quad f - f_0 \propto e^{i\omega t}, \tag{13.18}$$

where ω is the circular frequency. Thus $\partial/\partial t \to i\omega$, and *e.g.*

$$\frac{\partial(f - f_0)}{\partial t} = i\omega(f - f_0) = \frac{\partial f}{\partial t} \tag{13.19}$$

(we are only dealing with deviations from equilibrium). From Eq. (13.10) it follows that

$$\dot{\mathbf{v}} \cdot \nabla_v f = -\frac{f - f_0}{\tau} - i\omega(f - f_0) = -\frac{1}{\tau}[1 + i\omega\tau](f - f_0), \tag{13.20}$$

and hence

$$f - f_0 = -\frac{\tau}{1 + i\omega\tau} \dot{\mathbf{v}} \cdot \nabla_v f. \tag{13.21}$$

We see that $+i\omega\tau$ is the only new contribution. Thus for an oscillatory field F as in Eq. (13.18) we obtain by analogy with the first case

$$f - f_0 = -\frac{eF\tau}{m(1 + i\omega\tau)} \frac{v_x}{v} \frac{\partial f_0}{\partial v}. \tag{13.22}$$

13.3.2 Calculation of the current density

We consider below first the steady field. The current density j_x of electrons in the x-direction is the expression (recall the formula $j_x = \rho v_x$)

$$j_x = \sum_{\text{all possible states}} ev_x \frac{n}{V}, \tag{13.23}$$

where n is the number of electrons and V the volume. Since $n = gf$, we obtain

$$j_x = e \sum_{\text{states}} v_x f \frac{g}{V} = e \sum_{\text{states}} v_x(f - f_0) \frac{g}{V}. \tag{13.24}$$

Here we could replace f by $f - f_0$ since $j_x = 0$ for no external field. In the case of a perfectly isotropic system of electrons we have

$$\frac{g}{V} = 2 \frac{m^3 dv_x dv_y dv_z}{h^3},$$

and when averaged this does not contribute. Inserting from Eq. (13.17) $f - f_0$ into Eq. (13.24), we obtain

$$j_x = -\frac{e^2 F}{m} \sum_{\text{states}} \frac{\tau v_x^2}{v} \frac{\partial f_0}{\partial v} \left(\frac{g}{V}\right). \tag{13.25}$$

Averaging over all directions, we have

$$v^2 = v_x^2 + v_y^2 + v_z^2 = 3\overline{v_x^2}, \tag{13.26}$$

so that

$$j_x = -\frac{e^2 F}{3m} \sum_{\text{states}} \tau v \frac{\partial f_0}{\partial v} \left(\frac{g}{V}\right), \tag{13.27}$$

where when everything is isotropic (corresponds to integrating over all angles)

$$\left(\frac{g}{V}\right) = \frac{8\pi m^3 v^2 dv}{h^3}. \tag{13.28}$$

Hence we obtain

$$j_x = -\frac{8\pi e^2 m^2 F}{3h^3} \int \tau v^3 \frac{\partial f_0}{\partial v} dv. \tag{13.29}$$

The electrical conductivity σ is the ratio j_x/F (j_x, F both in x-direction). We obtain therefore — inserting the factor $(1 + i\omega\tau)$ for an oscillatory field — the following formula for the electrical conductivity:

$$\sigma = -\frac{8\pi e^2 m^2}{3h^3(1 + i\omega\tau)} \int \tau v^3 df_0. \tag{13.30}$$

13.3.3   Application to metals

In the Fermi–Dirac distribution function

$$f_0 = \frac{1}{e^{(\epsilon-\epsilon_0)/kT} + 1} \tag{13.31}$$

the quantity ϵ_0 is the energy when $v = v_0$, *i.e.* $\epsilon_0 = mv_0^2/2$. This energy ϵ_0 is also the energy at the Fermi surface as indicated in Fig. 13.1. Thus v_0 is

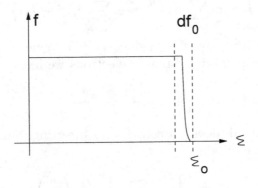

Fig. 13.1 The Fermi–Dirac distribution for a metal.

the velocity at the Fermi surface of the metal. In metals f_0 varies significantly only in the immediate vicinity of the velocity v_0, and hence df_0 is nonzero only in this region. We denote by τ_0 the value of τ at $\epsilon = \epsilon_0$. Then, assuming ϵ_0/kT to be very large,

$$\int df_0 = [f_0]_{\epsilon=0}^{\epsilon=\infty} = \left[\frac{1}{e^{(\epsilon-\epsilon_0)/kT}+1} \right]_{\epsilon=0}^{\epsilon=\infty} = 0 - \frac{1}{1+e^{-\epsilon_0/kT}} \simeq 0 - 1 = -1.$$

$$(13.32)$$

It follows that the electrical conductivity is given by

$$\sigma \simeq \frac{8\pi e^2 m^2 \tau_0 v_0^3}{3h^3}.$$

$$(13.33)$$

This expression can be seen to agree with that obtained semi-classically in electrodynamics. Let $\mathcal{N}$ be the product of the number of states ($= 8\pi m^3 v^2 dv/h^3$) and the probability of each state being occupied f_0; then the number of electrons per unit volume is

$$\frac{N}{V} = \mathcal{N} = \frac{8\pi m^3}{h^3} \int_0^\infty f_0 v^2 dv.$$

$$(13.34)$$

Here $f_0 = f_0(v)$. Therefore integrating by parts, we have

$$\int_0^\infty f_0 v^2 dv = \left[f_0 \frac{v^3}{3} \right]_0^\infty - \int_0^\infty \frac{df_0}{dv} \frac{v^3}{3} dv$$

$$= -\frac{1}{3} \int_0^\infty v^3 df_0 \simeq -\frac{1}{3} v_0^3(-1) = \frac{1}{3} v_0^3, \qquad (13.35)$$

since $f_0 = 0$ at $v = \infty$. Alternatively we could argue, $f_0 \simeq 1$ for v between 0 and v_0, so that the integral in Eq. (13.34) is approximately

$$\int f_0 v^2 dv \simeq 1. \int_0^{v_0} v^2 dv = \frac{1}{3} v_0^3.$$

$$(13.36)$$

It follows that

$$\mathcal{N} = \frac{8\pi m^3 v_0^3}{3h^3}, \quad \text{and} \quad \sigma = \frac{\mathcal{N} e^2 \tau_0}{m}.$$

$$(13.37)$$

This simple and useful formula for σ of a metal agrees with that of the simple semi-classical derivation given in texts of electrodynamics.[§] If the electric field is alternating the formula is (in units of (ohm · meter)$^{-1}$):

$$\sigma = \frac{\mathcal{N} e^2 \tau_0}{m(1+i\omega\tau_0)}.$$

$$(13.38)$$

[§]See [49], Eq. (12.22), p.338, and [50], Eq. (12.21), p.271.

For further discussion see Wannier [83], p.440. In the case of semiconductors the distribution function is rounded all the way, and hence the result is different (see Shockley [70], p.236). For further related discussion we refer to literature.[¶]

13.3.4   Calculation of the relaxation time

Our next step is the calculation of the collision or relaxation time τ in various contexts. We have seen that

$$\frac{\partial(f - f_0)}{\partial t} = -\frac{f - f_0}{\tau}. \tag{13.39}$$

Thus we can define the reciprocal of τ by the relation

$$\frac{1}{\tau} = -\frac{\partial(f - f_0)/\partial t}{(f - f_0)}. \tag{13.40}$$

We use this relation as our definition. Since from Eq. (13.24)

$$j_x = e \sum_{\text{states}} v_x(f - f_0)\frac{g}{V}, \tag{13.41}$$

where V is the volume, we see that $(f - f_0)$ is proportional to the current contribution in one direction from electrons of a given thermal velocity v. Therefore we can interpret our definition (13.40) by expressing:

$$\frac{1}{\tau(v)} = \begin{array}{l}\textit{relative decrease per unit time in the} \\ \text{current contribution in one direction.}\end{array} \tag{13.42}$$

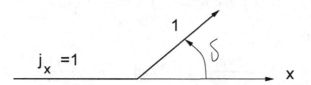

Fig. 13.2 Unit current contribution deflected in direction δ.

Next we consider a single collision. Moreover, we deal only with *elastic collisions* — for which the direction of motion changes but the kinetic energy and therefore the speed v remains unaltered as indicated in Fig. 13.2. Taking j_x before the collision as equal to 1, j_x after the collision is $\cos \delta$. Therefore

---

[¶]P.C. Riedl [62], pp.192-197.

the decrease in j_x is $1 - \cos \delta$, and the *relative decrease* in j_x is $(1 - \cos \delta)/1 = 1 - \cos \delta$. Taking n as the number of scattering elements per unit volume (*i.e.* from which an electron is scattered), v as the velocity of the incoming electron, and $P(\delta) = d\sigma/d\Omega$ as the probability that there shall be scattering in the direction $\Omega \to \delta$, we have:

number of collisions per unit time $= nvP(\delta)$ per unit solid angle Ω.
$$(13.43)$$

The expression for $P(\delta)$ is derived or given in texts on quantum mechanics for a spherically symmetric potential $\mathcal{V}(r)$ and in the Born approximation of the scattering amplitude.[||] We leave the derivation here to Example 13.1. One obtains

$$P(\delta) = |f^{\text{Born}}(E, E')|^2 = \left\{ \frac{2m}{\hbar^2 \mathcal{K}} \int_0^\infty \mathcal{V}(r) \sin \mathcal{K} r . r dr \right\}^2. \qquad (13.44)$$

For one electron incident per unit time per unit area, this gives the number of particles (electrons) scattered per unit time into unit solid angle, the angle of deviation being δ and

$$\mathcal{K} = 2k \sin \left(\frac{\delta}{2} \right), \qquad k^2 = \frac{2mE}{\hbar^2}. \qquad (13.45)$$

As shown in Example 13.1, this expression gives also the change in momentum $\hbar \mathcal{K}$ of the electron in the collision (k here not to be confused with the Boltzmann constant k). It follows that if scattering in any direction δ is taken into account (implying integration over $d\Omega$, *cf.* Fig. 13.3) we have

$$\frac{1}{\tau(v)} = nv \int_0^\pi P(\delta)(1 - \cos \delta) \underbrace{2\pi \sin \delta \, d\delta}_{d\Omega}, \qquad (13.46)$$

where $P(\delta) = d\sigma/d\Omega$ is known as the differential cross section, σ being the (total) cross section.

Scattering may ensue by various mechanisms. If the scattering is not too large, the probabilities of scattering by various mechanisms all add. Thus the $(1/\tau)'s$ add:

$$\frac{1}{\tau_{\text{total}}} = \frac{1}{\tau_{\text{1st mech.}}} + \frac{1}{\tau_{\text{2nd mech.}}} + \cdots . \qquad (13.47)$$

We consider several types of scattering.

[||] See H.J.W. Müller–Kirsten [46], Example 11.11, p.272, also Eq. (10.136), p.218, or in H.J.W. Müller–Kirsten [47], Eq. (10.136), p.198.

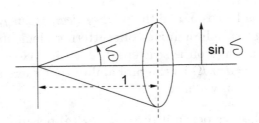

Fig. 13.3 The scattering angle δ.

   (i) *Scattering due to impurities* (only approximate in any case). We consider an impurity atom in a solid as indicated in Fig. 13.4.

We take the deviation in the potential due to the impurity atom to be localized within one atomic sphere as indicated in Fig. 13.4. The maximum value of r is the radius of the atomic sphere. Then if we are either dealing only with $\sin(\delta/2)$ small, or with k small as in semiconductors (electrons of low energy), we have

$$\mathcal{K}r \quad \text{small} \quad \longrightarrow \quad \sin \mathcal{K}r \quad \simeq \quad \mathcal{K}r. \tag{13.48}$$

Then the $\mathcal{K}$'s in Eq. (13.44) cancel in $P(\delta)$, and $P(\delta)$ is independent of the velocity (*i.e.* of k) of the electron. From Eq. (13.46) it follows that

$$\frac{1}{\tau} \propto v, \qquad \tau v = \text{const.} \tag{13.49}$$

Here τv is the product of the mean time between collisions with the velocity of the electron, and this is the *mean free path* of the electron. Thus the mean free path is approximately constant for impurity scattering. This is always true for semiconductors, but only approximately.

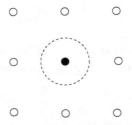

Fig. 13.4 The impurity atom and its atomic sphere.

   (ii) *Scattering by ionized impurity atoms in semiconductors.* This type of scattering of an electron from an impurity atom is provided for instance by a phosphorus ion P^+ in the position of a lattice of neutral silicon atoms Si as indicated in Fig. 13.5.

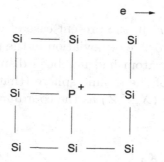

Fig. 13.5 The silicon lattice with impurity P^+.

In this case the disturbing radial potential $\mathcal{V}(r)$ is approximately given by the Coulomb potential, *i.e.*

$$\mathcal{V}(r) \simeq \frac{e^2}{\epsilon r}, \tag{13.50}$$

where ϵ is the dielectric constant of Si. At large distances the cloud of electrons surrounding the impurity atom has a screening effect so that $\mathcal{V}(r)$ decreases more rapidly at large distances than the Coulomb expression implies (see Example 13.2). Thus here we have a situation as in the case of the Rutherford scattering law. We know from the Rutherford scattering law derived in books on quantum mechanics** that ($f =$ scattering amplitude)

$$\frac{d\sigma}{d\Omega} = |f|^2 \propto \frac{1}{k^4} \propto \frac{1}{v^4}, \tag{13.51}$$

so that from Eq. (13.46) we obtain for the relaxation time τ the relation

$$\frac{1}{\tau(v)} \propto \frac{v}{\epsilon^2 v^4} = \frac{1}{\epsilon^2 v^3}. \tag{13.52}$$

(iii) *Thermal agitation at room temperature.* At room temperature T we can use classical statistics. Thus we can use the law of equipartition of energy of Sec. 4.3.6, and in particular Eq. (4.26b). The mean potential energy of an atom of mass m vibrating with natural frequency of vibration ν is (*cf.* Eq. (3.10))

$$\bar{\epsilon}_x = 2\pi^2 m \nu^2 \overline{x^2} = \frac{1}{2}kT, \tag{13.53}$$

**See *e.g.* [47], p.238, p.523, or [46], p.261, p.631.

where x is the displacement of the vibrating atom. We have therefore

$$\overline{x^2} \propto \frac{T}{m}. \tag{13.54}$$

What is the potential $\mathcal{V}(r)$ in the present case of a vibrating atom? We interpret $\mathcal{V}(r)$ as the disturbance or deviation in the potential from that which would have existed if the atom had not been displaced. We measure coordinates from the center of the atomic sphere (that around the undisturbed lattice position) and take (X, Y, Z) as the coordinates of the displaced atom

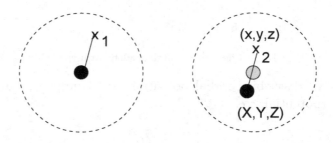

Fig. 13.6 The potentials at x_1, x_2 are assumed to be the same
(coordinates relative center of atomic sphere).

or nucleus. We assume that within one atomic sphere the disturbance potential $\mathcal{V}(r)$ at a point $\mathbf{r}$ a given distance $|\mathbf{r} - \mathbf{R}|$ from a displaced nucleus at $\mathbf{R}$ is the same as the potential for that same distance if the nucleus had not been displaced (this assumption is known as *Nordheim's assumption*). This point is illustrated in Fig. 13.6. We have (with the above interpretation)

$$\mathcal{V}(r) = V(x - X, y - Y, z - Z) - V(x, y, z)$$
$$\simeq -\left\{ X \frac{\partial V}{\partial x} + Y \frac{\partial V}{\partial y} + Z \frac{\partial V}{\partial z} \right\}, \tag{13.55a}$$

or

$$\mathcal{V}(r) = V(\mathbf{r} - \mathbf{R}) - V(\mathbf{r}) \simeq -\mathbf{R} \cdot \boldsymbol{\nabla}_r V(\mathbf{r}), \tag{13.55b}$$

so that $\mathcal{V}(r)$ is proportional to the displacement R of the atom and

$$\mathcal{V}^2(r) \propto \overline{R^2} \propto \frac{T}{m}, \tag{13.56}$$

where the last proportionality follows from Eq. (13.54). Since $P(\delta) \propto \mathcal{V}^2(r)$, we conclude that

$$P(\delta) \propto \frac{T}{m}. \tag{13.57}$$

If as for the impurity case (i), *i.e.* for semiconductors, $P(\delta)$ is independent of k, then Eq. (13.46) tells us that

$$\frac{1}{\tau} \propto \frac{T}{m}, \qquad \tau \propto \frac{m}{T}. \tag{13.58}$$

In electrodynamics it is shown that the relaxation time, *e.g.* of a dielectric, is proportional to the reciprocal of the resistivity.[††] Hence in the present case the resistance is proportional to the temperature T.

(iv) *Thermal agitation at very low temperatures.* At very low temperatures T classical statistics is inapplicable. However, as in Debye's theory of specific heat (*cf.* Chapter 9, there Eq. (9.5) and the earlier Eq. (8.14)), we have for the average of the square of the displacement X of an atom vibrating with natural frequency ν the relation

$$\overline{X^2} \propto \frac{h\nu}{e^{h\nu/kT} - 1}. \tag{13.59}$$

This quantity is negligible unless $h\nu \lesssim kT$. In analogy to the expression for the momentum of a photon (*cf.* Eq. (8.4) or [50], Eq. (17.78), p.416, or [49], Eq. (17.166), p.489), the momentum of a phonon is given by the ratio of its energy $h\nu$ divided by the velocity of sound u, *i.e.* here

$$\text{phonon momentum} = \frac{h\nu}{u} \lesssim \frac{kT}{u}, \tag{13.60}$$

where k is Boltzmann's constant. In the collision of a phonon with an electron, the change in momentum of the electron is $\hbar \mathcal{K}$, where $\mathcal{K}$ is given by Eq. (13.45). This change in momentum of the electron cannot be greater than the initial momentum of the phonon. Hence

$$\hbar \mathcal{K} = 2\hbar k_{\text{electron}} \sin\left(\frac{\delta}{2}\right) \lesssim \frac{k_{\text{Boltzmann}} T}{u}. \tag{13.61}$$

Hence, since we have small deviations δ, we obtain

$$\delta \propto T. \tag{13.62}$$

Thus in Eq. (13.46) the factor $(1 - \cos \delta) \simeq O(\delta^2)$ contains the temperature dependence T^2. Two more powers of T come from $\sin \delta d\delta$. Finally one further power of T comes from $P(\delta)$ as in Eq. (13.57) for those electrons for which $h\nu \lesssim kT$. Altogether we obtain therefore the temperature dependence

$$\frac{1}{\tau} \propto T^5. \tag{13.63}$$

[††]See *e.g.* [49], p.122, [50], p.112.

Thus at very low temperatures $1/\tau \propto T^5$, and thus the resistance of the solid is also proportional to T^5. This temperature dependence is called the *Bloch T^5 law*.

13.4   Applications and Examples

Example 13.1: Probability of scattering in direction δ
Derive formula (13.44).

Solution: The total elastic cross section σ_{tot} is defined in such a way that classically it is just the equivalent area of obstruction. One therefore defines σ_{tot} as the total number of particles scattered per unit time in all directions for one particle incident per unit time per unit area, or

$$\sigma_{\text{tot}} = \int \frac{d\sigma}{d\Omega} d\Omega, \tag{13.64}$$

where σ is the number of particles scattered per unit time into unit solid angle in the direction of Ω for one particle incident per unit time per unit area. The differential elastic cross section $d\sigma/d\Omega$ is the modulus squared of the scattering amplitude $f(\Omega)$, or in *Born approximation*,

$$\frac{d\sigma}{d\Omega} = |f^{\text{Born}}(\mathbf{k}, \mathbf{k}')|^2, \tag{13.65}$$

where (with $\mathbf{k}^2 = \mathbf{k}'^2$ for elastic scattering)

$$f^{\text{Born}} = -\frac{m}{2\pi\hbar^2} \int e^{i(\mathbf{k}-\mathbf{k}')\cdot\mathbf{x}'} V(\mathbf{x}')d\mathbf{x}', \tag{13.66}$$

and the scattering angle δ is contained in the expression

$$\mathcal{K} = |\mathbf{k} - \mathbf{k}'| = \sqrt{(\mathbf{k} - \mathbf{k}')^2} = \sqrt{2k^2 - 2k^2 \cos\delta}$$
$$= \sqrt{2k^2(1 - \cos\delta)} = \sqrt{2k^2 \cdot 2\sin^2 \delta/2} = 2k\sin(\delta/2). \tag{13.67}$$

It follows that for a spherically symmetric potential $\mathcal{V}(r)$ one obtains by integration over the angles Eq. (13.44), *i.e.*

$$f^{\text{Born}} = -\frac{2m}{\hbar^2 \mathcal{K}} \int_0^\infty r\,dr\,\mathcal{V}(r)\sin\mathcal{K}r. \tag{13.68}$$

13.5   Problems without Worked Solutions

Example 13.2: Screening effect of the electron cloud
Consider the scattering of an electron by an ionized impurity atom in a semiconductor (see Sec. 13.3.4, case (ii)). Assume an exponential screening (due to the electron cloud) of the Coulomb potential and calculate the relaxation time τ.
[Hint: One can evaluate the integral

$$\int_0^\infty \sin\alpha r\,e^{-\beta r}\,dr = \frac{\alpha}{\beta^2 + \alpha^2}, \quad \beta > 0, \tag{13.69}$$

either by expressing $\sin\alpha r$ in terms of exponentials, or by looking up a Table of Laplace Transforms].[‡‡]

[‡‡]See *e.g.* W. Magnus and F. Oberhettinger [40], p.125.

Example 13.3: Maxwell–Boltzmann distribution

Show that in the presence of an external potential $U(\mathbf{r})$ the stationary solution of the Boltzmann equation is the Maxwell–Boltzmann distribution.[§§]

Example 13.4: The insulator versus superconductor

Why is an insulator not a perfect superconductor?

[Hint: Recall Eq. (13.23). If the conduction band is full, the velocity of the electrons at the upper edge is zero. No current is induced by a magnetic field in a completely filled conduction band. The kinetic energy has a point of inflection and hence a horizontal tangent at the upper end].[¶¶]

[§§]See, if you like, Ya.P. Terletskii [75], pp.167, 255.

[¶¶]V.F. Weisskopf [84].

Chapter 14

Thermal Radiation of Black Holes

14.1 Preliminary Remarks

In Chapter 8 we considered thermal radiation of a black body. A familiar concept which comes to one's mind in this context is that of a black hole in cosmology. It is well-known that this term arises in Einstein's general theory of relativity and in the work of S. Hawking, and is surrounded with an aura of mystification. Although this topic lies outside our perspectives here, we can venture a glimpse into it, and at least gain an idea of its relation to the thermal radiation treated in earlier chapters. Thermal radiation means heat and this implies temperature. How does temperature arise in the context of black holes, and what are these black holes — how are they related to the black bodies considered in Chapter 8? These are the type of questions we attempt to address in the following.

14.2 Background Geometry

In the massless 4-dimensional Minkowski spacetime with coordinates t, x, y, z the element of a distance, ds, is given by the metric equation

$$ds^2 = -c^2 dt^2 + (dx^2 + dy^2 + dz^2) \equiv \Sigma_\mu dx_\mu dx^\mu, \qquad (14.1)$$

where c is the velocity of light. In the presence of a mass M as the source of gravity the gravitational field appears in the coefficients $g_{\mu\nu}$, *i.e.* in the 4-dimensional Minkowski space with line-element ds given by

$$ds^2 = \Sigma_{\mu\nu} g_{\mu\nu} dx^\mu dx^\nu. \qquad (14.2)$$

The tensor field components $g_{\mu\nu}$, not all equal to plus or minus 1, depend on M and are obtained as solutions of Einstein's equation. One can choose various types of coordinate systems.

The simplest case of a black hole is the spherically symmetric static uncharged black hole — imagined as a spherical region of otherwise empty space described by the corresponding solution of the Einstein equation in general relativity. This is given in spherical Schwarzschild coordinates ct, r, θ, ϕ by the metric equation

$$ds^2 = \Sigma_{\mu\nu} g_{\mu\nu} dx^\mu dx^\nu$$

$$= -c^2 \left(1 - \frac{2MG}{c^2 r} \right) dt^2 + \left(1 - \frac{2MG}{c^2 r} \right)^{-1} dr^2 + r^2 d\Omega^2, \quad (14.3)$$

with $d\Omega^2 = d\theta^2 + \sin^2\theta d\phi^2$ (for a detailed derivation see [48], Chapter 16). Here M is the mass of the source of the gravitational field (which is described by the coefficients of the metric equation), c is again the velocity of light, and G is the gravitational constant with value $G = 6.67 \times 10^{-11} \mathrm{m}^3 \mathrm{kg}^{-1} \mathrm{s}^{-2}$. Thus Newton's potential would be given by $\phi(r) = -GM/r$ per unit mass. The quantity $r_S = 2MG/c^2$ is called the Schwarzschild radius, and the spherical area at this distance is called the event horizon. In general the mass M could be the mass of the Earth or that of the sun or of something else. In our context here it is the mass of a black hole (with nothing else around).

The above Schwarzschild metric is seen to satisfy the asymptotic condition of a flat space at $r = \infty$. The singularity at $r = r_S$, called the event horizon of the black hole — which is a light-like surface — is not a singularity of the metric but of the system of coordinates (however, $r = 0$ is a true local singularity) — it can be avoided by changing to a different system of coordinates. The horizon has no extension in the direction of time t (i.e. in g_{00}) and is therefore described as static. The radius r_S is the radius of the macroscopic body — in our case that of the interior of the black hole where the Schwarzschild coordinates are invalid — hence the description as black — radiation or matter can pass into it, but cannot escape. This eternal black hole as a spherical shell of gravitating matter of mass M is an idealization (real black holes result from collapse of gravitating matter). A traveller crossing the horizon will not notice anything but his observer far outside cannot receive a signal sent to him from inside the shell.

14.3   Rindler Coordinates

In order to learn more about the domain just outside the static black hole or shell and near the horizon, it is convenient to use more appropriate coordinates. These coordinates, called Rindler coordinates, ω, ρ, X, Y, are such that ρ measures the proper radial distance from the horizon, and X and Y replace the angular variables. One also introduces the dimensionless Rindler time ω by setting

$$\omega = \frac{c^3 t}{4MG}. \tag{14.4}$$

The transformation of the 3 spatial Schwarzschild coordinates to Rindler coordinates is given by*

$$\rho = \int_{2MG/c^2}^{r} \frac{dr'}{\sqrt{1 - \frac{2MG}{c^2 r'}}}, \quad \begin{pmatrix} X \\ Y \end{pmatrix} = \frac{2MG\theta}{c^2} \begin{pmatrix} \cos\phi \\ \sin\phi \end{pmatrix}, \tag{14.5}$$

where ρ measures the proper outside radial distance from the horizon. We leave the transformation of the Schwarzschild metric equation ds^2 as an exercise or as consultation of the literature [73]. The result (with near horizon approximation) is the Rindler metric equation

$$ds^2 = -\rho^2 d\omega^2 + d\rho^2 + dX^2 + dY^2, \tag{14.6}$$

In Example 14.1 it is shown that the distant observer at proper distance ρ far away from the black hole experiences an acceleration a of the order of $1/\rho$. In Example 14.2 it is shown that the proper time τ of the far-away observer is related to ρ such that

$$c\tau = \rho\omega. \tag{14.7}$$

With another transformation

$$cT = \rho \sinh(\omega), \quad Z = \rho \cosh(\omega), \tag{14.8}$$

this metric becomes of familiar Minkowski type:

$$ds^2 = -c^2 dT^2 + dZ^2 + dX^2 + dY^2. \tag{14.9}$$

Thus locally the horizon does not lead to a singularity and — again locally — it is almost a flat spacetime. An observer A permanently stationed far outside the black hole can observe a particle falling radially into the black hole. Another observer B following the particle cannot communicate with A once he crossed the horizon. The horizon is a global property, not a local one. In Example 14.2 it is shown that — ignoring coordinates X and Y and considering the transformation from the 2-dimensional Minkowski space to

---

*See L. Susskind and J. Lindesay [73], p.8.

Rindler space — the observer at fixed ρ traces a hyperbola in Minkowski space. With the further substitution

$$u = \ln \rho, \qquad \rho = e^u \times \text{unit length}, \qquad (14.10)$$

(implying $u = -\infty$ at the horizon where $\rho = 0$), the metric becomes

$$ds^2 = -e^{2u}[d\omega^2 - du^2] + dX^2 + dY^2. \qquad (14.11)$$

This metric will be referred to below.

The metric equation above describes the deformation of ordinary space-time due to the presence of the mass M — here taken to be that of a 3-dimensional spherical shell. Hence this metric provides the geometry of the space which is relevant here.

14.4   Introduction of Fields

In order to obtain a theory which involves radiation, we have to introduce one or more types of fields, like a scalar field or a vector field (like that in electrodynamics) in this space, *i.e.* in the neighbourhood of the shell of mass M. The field easiest to handle is a massless scalar field $\chi(x_\mu)$ (also known as Klein–Gordon field) which suffices in any case for our purposes here. This is expressed by saying, one introduces the scalar field in the background of the gravitational field contained in the above Schwarzschild metric (in the coefficients $g_{\mu\nu}$).

The general procedure for developing such a theory is to define a Lagrange functional L in a way very analogous to what one does in classical mechanics (*cf.* Eq. (8.6)). Since we consider fields, we consider (as in electrodynamics) field densities in a volume V. The appropriate integral over the spacetime volume is called action — and this is varied to derive the equations of motion like the Euler–Lagrange equations in mechanics. It is illuminating to hear what R. Penrose says about this procedure. In his book *The Road to Reality, A complete guide to the laws of the universe*, pp.489-492, [55] he says: "Lagrangian theory (as well as Hamiltonian theory) has a highly influential role in modern physics ... when some suggested new theory is put forward, it is almost invariably given in the form of some Lagrangian functional ...".

The action for the scalar field χ is given by the integral

$$I = \frac{1}{2} \int d^4x \sqrt{-g} g^{\mu\nu} \partial_\mu \chi \partial_\nu \chi \equiv \int d^4x \sqrt{-g} L, \qquad (14.12)$$

where $g = \det(g_{\mu\nu})$ (in ordinary Minkowski spacetime this is simply -1, the product of the diagonal elements) and the components $g^{\mu\nu}$ are given by the metric equation. The square root of $-g$ is necessary to ensure invariance of the volume element under Lorentz transformations.[†] In the case of the Rindler metric (obtained from the original Schwarzschild metric) the action becomes in units with $c = 1$ (note the change from ρ to u above)[‡]

$$I = \frac{1}{2} \int dX\, dY\, du\, d\omega \left[\left(\frac{\partial \chi}{\partial \omega} \right)^2 - \left(\frac{\partial \chi}{\partial u} \right)^2 - e^{2u}(\partial_\perp \chi)^2 \right], \qquad (14.13)$$

where $\partial_\perp \chi = (\partial/\partial X, \partial/\partial Y)$. The corresponding Hamiltonian $H_R \propto H$ can be derived from I in the usual way with a Legendre transform as explained in Sec. 1.2. Decomposing the field χ into transverse plane waves χ_k (a Fourier transform) with transverse wave vector $k_\perp$, $k_\perp^2 = k_X^2 + k_Y^2 \equiv k^2$, we obtain for a given wave number k (hence no k-integration in the following)

$$I_k = \frac{1}{2} \int du\, d\omega \left[\left(\frac{\partial \chi_k}{\partial \omega} \right)^2 - \left(\frac{\partial \chi_k}{\partial u} \right)^2 - k^2 e^{2u} \chi_k^2 \right]. \qquad (14.14)$$

By also setting

$$\chi_k(u, \omega) = e^{i\lambda\omega} \tilde{\chi}_k(u),$$

the Euler–Lagrange equation of motion obtained from I_k is the Schrödinger-like wave equation (again in units with $c = 1$)

$$-\frac{\partial^2 \tilde{\chi}_k(u)}{\partial u^2} + k^2 e^{2u} \tilde{\chi}_k(u) = \lambda^2 \tilde{\chi}_k(u). \qquad (14.15)$$

Since $u = \ln \rho$, u is $-\infty$ at the horizon where $\rho = 0$, and one can conclude that the exponential potential in Eq. (14.15) confines quanta of the scalar field χ_k near the horizon. This is a simplified but qualitatively satisfying conclusion which serves our purposes here. However, as shown in the book of L. Susskind and J. Lindesay [73], if one looks at the boundary conditions and the quantum mechanics more closely, there are grave problems which we cannot enter into here. However, the entropy stored in the field χ_k in Rindler space can be estimated from the density of a $1+1$ dimensional massless free boson at temperature T — see Example 9.10 and [73], p.47.

Before we proceed we recall (now with subscripts) the relation between Schwarzschild time t_S and (dimensionless) Rindler time $\omega = t_R$, i.e. (see Eq. (14.4))

$$ct_S = \frac{4MG}{c^2} t_R, \qquad (14.16)$$

[†]For explicit verification see [48], Secs. 6.4.3 and 14.4.3.
[‡]See L. Susskind and J. Lindesay [73], p.28

which has the consequence that the Rindler frequency ν_R of a quantum field is seen by a far-away observer as

$$\nu_S = \frac{c^3}{4MG}\nu_R. \tag{14.17}$$

This relation implies that the frequency ν_S observed by the far-away observer appears reduced, *i.e.* as "red shifted" as one says.[§] This is similar to the Doppler effect.[¶] A high frequency whistle that flies (*i.e.* falls) towards the horizon becomes audible to the far-away observer at a red-shifted frequency.

14.5   Thermalization

Quantum field theory differs from classical field theory (like classical electro-dynamics) in that it allows the creation and annihilation of particles (in quantum electrodynamics, for instance, the creation of electron-positron pairs by vacuum polarization). Since such field theories have to be relativistic (formulated in terms of operators in a Hilbert space, the space of states), real processes take place in the time-like region of space (the forward light-cone), *i.e.* in the region in which signals can be sent or received, and equal-time operators at different spatial points do not commute (receiving a signal from a space-like point would require a velocity larger than c, *i.e.* from outside the light-cone). In a state of equilibrium we can imagine this state called vacuum to allow a constant creation and annihilation of such virtual pairs in short periods of time. Since we began with Minkowski space we have to relate light-cone considerations there to those of Rindler space (outside the black hole horizon), and to observations made by observers A and B above.

In quantum mechanics and in classical field theory the evolution of one surface of constant time to another is governed by the Hamiltonian H (Hamilton or boost operator or time-shift operator) as described by the time-dependent Schrödinger equation

$$i\hbar\frac{\partial\Psi}{\partial t} = H\Psi, \quad \Psi_t = \Psi_0\exp(Ht/i\hbar). \tag{14.18}$$

Similarly the Rindler Hamiltonian is the operator which shifts proper time ω from one surface of constant ρ to another. Equation (14.7) shows that this shift is ρ-dependent.

The state with lowest eigenvalue of H is the ground state in quantum mechanics. In quantum field theory the lowest state is the vacuum — and

---

[§]For a detailed explanation of this term see [48], Sec. 13.5.
[¶]See *e.g.* [48], p.421.

that is not empty space but visualized as a sea of virtual particles and antiparticles with their fields at every point in Minkowski space. These fields are correlated or entangled with one another in the sense of being able to send signals to one another in the forward light-cone.

At any particular time a system may therefore be in one of a number of admittable states as a result of correlation with the rest of the universe. This system of states is described by the von Neumann density matrix ρ_N (*i.e.* operator in matrix representation) which satisfies an equation analogous to the Schrödinger equation above[‖], *i.e.*

$$\frac{\partial \rho_N}{\partial \beta} = -H\rho_N, \quad \rho_N = \frac{\exp(-\beta H)}{\text{Tr}\exp(-\beta H))}. \tag{14.19}$$

Here the trace in the denominator is the normalization factor. The parameter β assumes the meaning of inverse temperature, *i.e.* $\beta = 1/kT$, as a result of the description of a system (as embedded in the rest of the world) by the appropriate matrix expectation value of ρ_N.

Thus the density matrix ρ_N describes one or more systems in contact with other systems in an overall enclosure or universe, and therefore describes a canonical ensemble (a temperature cannot be assigned to a single particle) with a mixing of states (*i.e.* several systems with — in general — different weightings). Comparison of the Schrödinger equation (14.18) with the density equation (14.19) already hints in the direction of a correspondence between time in ordinary field theory and inverse temperature in the thermalized field theory. In fact, that is what the arguments below are leading to.

In Minkowski spacetime we have the freedom to change t into it, *i.e.* to go to imaginary time, in which case Lorentz invariance in Minkowski spacetime would become rotational invariance in a 4-dimensional Euclidean spacetime. In order to maintain the relation between Schwarzschild time and Rindler time we have to replace ω by $i\omega$ and hence the Rindler metric equation (14.6) becomes a polar-coordinate-like metric, or the invariance under ω translations becomes invariance under rotations in Euclidean spacetime, the hyperbolic angle ω becoming a real angle. In order to cover the entire angular space, rotations must close after a rotation of 2π in the Euclidean metric, and only an angle of 2π is relevant. Hence

$$e^{Ht/i\hbar} \longrightarrow e^{-\beta_R H_R} \longrightarrow e^{-2\pi H_R}. \tag{14.20}$$

Thus the period of imaginary time corresponds to temperature in thermal field theory. The rigorous justification of this step is given in L. Susskind

---

[‖]See R.P. Feynman [25].

and J. Lindesay, Chapter 3 [73]. The Rindler temperature T_R is therefore given by $\beta_R = 2\pi$. This T_R is not a real temperature, as we see also from the fact that it is dimensionless like H_R. The arguments leading to this result are of wide generality and were first given by W.G. Unruh [79].

The temperature T in terms of the parameter $\beta = 1/kT$ arose in the present context only with the introduction of the density matrix ρ_N. Comparison of the Schrödinger equation (14.18) (which describes a pure state) with the density or von Neumann equation (14.19) (for a system of mixed states) shows that in proceeding from the first to the other $Ht/i\hbar$ is replaced by $-\beta H$. Thus if we pass to observer A's proper time τ by setting $t = -i\tau$, and change with Eq. (14.7), *i.e.* $c\tau = \rho\omega$, to the Rindler frame, the quantity $Ht/i\hbar$ becomes $-H\tau/\hbar$ and this becomes $-(\rho/\hbar c)\omega H$. This is thermalized by equating it to $-\beta H$ in the density exponential. By going to imaginary time, the original Rindler time ω became an angle β_R, the product of Hamiltonian times β being the same, and so (β has dimension of inverse energy, β_R is dimensionless)

$$\beta = \frac{\rho}{\hbar c}\beta_R \qquad (14.21)$$

is the relation between the temperatures observed by A and B. The temperature T of observer A is therefore given by (in degrees Kelvin)

$$T = \frac{\hbar c}{k\rho}\frac{1}{\beta_R} = \frac{a\hbar}{kc}\frac{1}{\beta_R}, \qquad (14.22)$$

where a is the uniform acceleration of the observer (see after Eq. (14.6) and Example 14.1). Thus the temperature observed by far-away observer A increases with decreasing ρ, *i.e.* in approaching the event horizon. This effect, observation of temperature by an accelerating observer (*cf.* Example 14.1), is called Unruh effect. Trivialized it says that an accelerated thermometer ought to show an increase in temperature due to its acceleration alone.

14.6 Black Hole Evaporation

In the above we considered the simplest type of field theory, *i.e.* that of a massless scalar field, in the background of a Schwarzschild black hole. In view of correlations between fields at equal times but different locations, different states and thermal fluctuations arise which, when in equilibrium, imply a definite temperature of the ensemble of mixed states. The thermal fluctuations are the vacuum fluctuations due to creation and annihilation of particle-antiparticle pairs in short periods of time in the region just outside the black hole. As L. Susskind and J. Lindesay [73] (p.41) say, the horizon behaves like a hot membrane radiating and absorbing thermal energy, *i.e.* the particles there carry off energy in the form of thermal radiation. We observed previously (*cf.* Eq. (14.17)) that the distant Schwarzschild observer has a red-shifted frequency ν_S compared with the Rindler frequency ν_R. Since dimensionally energy $E = h\nu$ and $E = kT = 1/\beta$, the temperature observed by the far-away observer is also red-shifted, *i.e.***

$$\beta_S = \frac{4MG}{\hbar c^3}\beta_R, \quad T_S = \frac{\hbar c^3}{8\pi MGk} = \frac{3.08 \times 10^{24}}{8\pi M}{}^{\circ}K. \tag{14.23}$$

This relation was first obtained by S. Hawking [30] and the temperature T_S is therefore called Hawking temperature. Since massless particles can also escape to infinity the black hole loses energy to its surroundings, but as it radiates its temperature increases. The black hole is therefore similar to a leaky cavity containing thermal radiation. The Hawking temperature differs from the Unruh temperature in that the acceleration of the observer in the case of the latter is the acceleration due to the surface gravity g of the black hole in the case of Hawking, *i.e.*

$$g = G\frac{1 \times M}{r_S^2} = \frac{c^4}{4MG}. \tag{14.24}$$

We have here an example illustrating the so-called strong form of Einstein's principle of equivalence.†† This principle says that the effect of a gravitational field can equivalently be studied by considering the same effect in an accelerated coordinate system, *i.e.* an observer cannot distinguish between the two.

14.7 Applications and Examples

Example 14.1: Acceleration a of a distant observer
Show that for $\rho \ll MG$ the acceleration a of an observer far away from the black hole is of the order of $1/\rho$.

Solution: A theorem by G.D. Birkhoff [5] states that the spherically symmetric spacetime around any spherically symmetric matter configuration has the same properties as spacetime around a Schwarzschild black hole of the same mass. We have therefore according to Newton: acceleration $\propto \partial V/\partial r$ with potential $V(r) \propto (2MG+\rho(r))^{-1}$. Then (with $\partial\rho/\partial r$ from the integral in Eq. (14.5) and $c = 1$)

$$\frac{\partial V}{\partial r} = \frac{\partial V}{\partial \rho}\frac{\partial \rho}{\partial r} \simeq \frac{1}{(2MG)^2}\frac{1}{\sqrt{1 - 2MG/r}} = \frac{\sqrt{r}}{(2MG)^2\sqrt{r - 2MG}} \approx \frac{1}{(2MG)^{3/2}\sqrt{r - 2MG}}. \tag{14.25}$$

**This is formula (4.1.25) in [73] with $\hbar = c = k = 1$.
††For a detailed explanation of this principle see [48], pp.359–363.

But from integration of Eq. (14.5)

$$\rho = \sqrt{r(r-2MG)} + 2MG\sinh^{-1}(\sqrt{r/2MG}-1) \simeq 2\sqrt{2MG(r-2MG)}, \qquad (14.26)$$

where we used from Mathematical Tables that $\sinh^{-1}x \simeq x$ for $x^2 < 1$. Hence acceleration $a \simeq 1/\rho$.

Example 14.2: Transformation to Rindler space

The transformation from 2-dimensional Minkowski space with coordinates ct, x to Rindler space with coordinates ω, ρ is given by the equations

$$x = \rho\cosh(\omega), \quad ct = \rho\sinh(\omega). \qquad (14.27)$$

The dimensionless coordinate ω is called Rindler time. Show that an observer moving with fixed parameter ρ traces out a hyperbola in Minkowski space. Show also that this observer's proper time τ is given by

$$\tau = \frac{\rho\omega}{c}. \qquad (14.28)$$

Solution: Using the formula $\cosh^2 x - \sinh^2 x = 1$, we obtain from the coordinate transformation the equation of a hyperbola in Minkowski space, *i.e.*

$$x^2 - c^2 t^2 = \rho^2. \qquad (14.29)$$

The proper time of the observer in Minkowski space is the time τ in his own coordinate system with line element ds^2 given by the metric equation

$$ds^2 = -c^2 dt^2 + dx^2 \quad \text{and hence} \quad d\tau^2 = -ds^2/c^2. \qquad (14.30)$$

The line element in Rindler space is given by the metric equation

$$ds^2 = -\rho^2 d\omega^2 + d\rho^2, \qquad (14.31)$$

where ω is the Rindler time. Thus proper Minkowski time τ transforms such that

$$c^2 d\tau^2 = \rho^2 d\omega^2 - d\rho^2. \qquad (14.32)$$

For $\rho = $ const. we have

$$cd\tau = \rho d\omega, \quad c\tau = \rho\omega, \quad d\omega/d\tau = c/\rho = \text{const.} \qquad (14.33)$$

14.8   Problems without Worked Solutions

Example 14.3: Unruh temperature

Show (using Example 14.1) that the proper acceleration a of the far-away observer is given by

$$a \simeq \frac{c^2}{\rho}, \qquad (14.34)$$

and hence the Unruh temperature by

$$T \simeq \frac{a\hbar}{2\pi kc}. \qquad (14.35)$$

What is the acceleration corresponding to 1 degree Kelvin?

Example 14.4: Entropy of a black hole

Use the second law of thermodynamics with energy E replaced by the mass of the black hole to show that the entropy S of the black hole is given by

$$S = \frac{4\pi G k}{\hbar c} M^2, \tag{14.36}$$

or, expressed in terms of the Schwarzschild radius $r_S = 2MG/c^2$,

$$S = \frac{kc^3}{4\hbar G} A, \tag{14.37}$$

where A is the area of the horizon. The last expression is known as the Bekenstein–Hawking entropy of a black hole. Different from ordinary thermodynamics — where entropy is proportional to volume — the entropy is here proportional to area. Since entropy counts the number of quantum states, this has the consequence that quantum gravity has vastly fewer quantum states than any ordinary quantum field theory.

Example 14.5: Stefan–Boltzmann law

Assuming that per unit Rindler time one quantum of the scalar field escapes from the horizon domain, the flux of escaping quanta is of the order of the Schwarzschild frequency, *i.e.* $c^3/4MG$ per second. Each quantum carries energy of the order of $h\nu_S \equiv kT_S$. The rate of energy loss per unit time is therefore

$$\frac{c^3}{4MG} kT_S,$$

which we equate to

$$\frac{dMc^2}{dt} = -L, \tag{14.38}$$

where L is called luminosity. Hence, with kT_S from Eq. (14.23),

$$\frac{dMc^2}{dt} = \frac{\hbar c^6}{2\pi(4MG)^2}. \tag{14.39}$$

Show that the luminosity per unit area is

$$\frac{1}{\text{area}} \frac{dMc^2}{dt} = \frac{k^4\pi^2}{c^2} T_S^4. \tag{14.40}$$

We recognise this result as the Stefan–Boltzmann law (see Chapter 8 and L. Susskind and J. Lindesay [73], pp.54, 143).

Example 14.6: Specific heat of a black hole

Use the definition of specific heat and $E = Mc^2$ to show that the specific heat of a black hole is negative. What does this imply?

Example 14.7: Mass of the black hole at $T_S = 1\text{GeV}$

What is the mass of the black hole at a temperature of 1GeV? The result is a tiny fraction of the original mass of the black hole of the order of $10^{57}\,\text{GeV}/c^2$.

[Hint: $1\text{GeV} = 1.78 \times 10^{-27}c^2\text{kg}, k = 1.38 \times 10^{-23}\,\text{J}/^\circ\text{K}$]
(Answer: $8 \times 10^9 kg$)

Bibliography

[1] W.P. Allis and M.A. Herlin, *Thermodynamics and Statistical Mechanics* (McGraw–Hill, 1952).

[2] D.J. Amit and Y. Verbin, *Statistical Physics, An Introductory Course* (World Scientific, 1999). This text is an introduction in the true sense of the word, and is therefore recommendable for beginners.

[3] M.H. Anderson, J.R. Ensher, M.R. Matthews, C.E. Wieman and E.A. Cornell, *Science* **269** (1995) 198.

[4] A. Ben–Naim, *Entropy Demystified* (World Scientific, 2007).

[5] G.D. Birkhoff, *Relativity and Modern Physics* (Harvard University Press, 1923).

[6] M.G. Bowler, *Lectures on Statistical Mechanics* (Pergamon Press, 1982).

[7] A.F. Brown, *Statistical Physics* (Edinburgh University Press, 1968).

[8] A.D. Buckingham, *The Laws and Applications of Thermodynamics* (Pergamon Press, 1964).

[9] K. Burnett, M. Edwards and C.W. Clark, The theory of Bose–Einstein condensation of dilute gases, *Phys. Today* **52** (1999) 37–42.

[10] H.B. Callen, *Thermodynamics and an Introduction to Thermostatics*, 2nd ed. (Wiley, 1985).

[11] C. Cercignani, *Theory and Application of the Boltzmann Equation* (Scottish Academic Press, 1975).

[12] D. Chandler, *Introduction to Modern Statistical Mechanics* (Oxford University Press, 1987).

[13] J. Clunie, *Proc. Phys. Soc.* **A67** (1954) 632.

[14] R.Courant, *Differential and Integral Calculus*, 2nd ed., Vols. I, II (Blackie, 1952).

[15] C.G. Darwin and R.H. Fowler, *Phil. Mag.* **44** (1922) 450 and 823.

[16] P. Debye, *Math. Ann.* **67** (1909) 535.

[17] R.B. Dingle, *Asymptotic Expansions: Their Derivation and Interpretation* (Academic Press, 1973). The derivation of the Bose–Einstein distribution under condensation conditions is treated on pp. 267–271. The book also discusses Bose–Einstein and Fermi–Dirac integrals.

[18] R.B. Dingle, Fermi–Dirac integrals, *Appl. Sci. Res.* **B6** (1957) 225-239.

[19] R.B. Dingle, Bose–Einstein integrals, *Appl. Sci. Res.* **B6** (1957) 240-244. For related work see R.B. Dingle, *Appl. Sci. Res.* **B4** (1955) 401 and R.B. Dingle, D. Arndt and S.K. Roy, *Appl. Sci. Res.* **B6** (1957) 144.

[20] R.B. Dingle, The Bose–Einstein statistics of particles with special reference to the case of low temperatures, *Proc. Camb. Phil. Soc.* **45** (1949) 275–287.

[21] R.B. Dingle, *Adv. Phys.* **1** (1952) 111–168. Section 3.2 of this paper deals with many aspects of Bose–Einstein condensation.

[22] J.J. Duderstadt and W.R. Martin, *Transport Theory* (John Wiley, 1979).

[23] H.B. Dwight, *Tables of Integrals and other Mathematical Data*, 3rd ed. (Macmillan, 1957).

[24] E. Fermi, *Nuclear Physics* (The University of Chicago Press, 1950).

[25] R.P. Feynman, *Statistical Mechanics* (W.A. Benjamin, 1972).

[26] T. Fliessbach, *Statistische Physik*, 2nd ed. (Spektrum Akad. Verlag, 1995).

[27] R.H. Fowler, *Statistical Mechanics*, 2nd ed. (Cambridge University Press, 1952). This book employs the Darwin–Fowler method.

[28] R.P.H. Gasser and W.G. Richards, *An Introduction to Statistical Thermodynamics* (World Scientific, 1995). This text is brief but very readable, and recommendable for beginners.

[29] S. Greenspoon and R.K. Pathria, *Phys. Rev.* **A9** (1974) 2103.

[30] S.W. Hawking, Black hole explosions?, *Nature* **248** (1974) 30-31.

[31] H.J.G. Hayman, *Statistical Thermodynamics* (Elsevier Pub. Co., 1967).

[32] W. Heitler, *Proc. Camb. Phil. Soc.* **32** (1936) 112.

[33] T.L. Hill, *Statistical Mechanics* (MacGraw-Hill, 1956).

[34] T.L. Hill, *An Introduction to Statistical Thermodynamics* (Dover Publ., 1986).

[35] K. Huang, *Introduction to Statistical Physics* (Taylor & Francis, 2001). This highly recommendable book treats also applications, such as the order parameter, superfluidity, noise and stochastic processes.

[36] J.H. Jeans, *Introduction to the Kinetic Theory of Gases* (Cambridge Univ. Press, 1940).

[37] D. Kleppner, The fuss about Bose–Einstein condensation, *Phys. Today* **49**(8) (1996) 11.

[38] P.T. Landsberg, On Bose–Einstein condensation, *Proc. Camb. Phil. Soc.* **50** (1954) 65–76. This paper compares and evaluates various rigorous discussions of Bose–Einstein condensation. It is also stated that the phenomenon does not occur in lower than three spatial dimensions, and considers only systems with an infinity of quantum states (a system having a finite number of quantum states with N/V finite and nonzero being condensed at all temperatures).

[39] P.T. Landsberg, *Thermodynamics and Statistical Mechanics* (Oxford Univ. Press, 1978). This detailed book contains a section with worked solutions to problems in the text.

[40] W. Magnus and F. Oberhettinger, *Formulas and Theorems for the Functions of Mathematical Physics* (Chelsea Publ. Co., 1954).

[41] F. Mandl, *Statistical Physics* (John Wiley, 1974).

[42] Ma, Shang–Keng, *Statistical Mechanics* (World Scientific, 1985).

[43] A. Messiah, *Mécanique Quantique*, Vol. 1 (Dunod, 1969), *Quantenmechanik*, translated by J. Streubel (de Gruyter, 1976).

[44] D.C. Mattis and R.H. Swendsen, *Statistical Mechanics made Simple*, 2nd ed. (World Scientific, 2008). This is an advanced text.

[45] J.E. Mayer and M.G. Mayer, *Statistical Mechanics*, 2nd ed. (John Wiley, 1977).

[46] H.J.W. Müller–Kirsten, *Introduction to Quantum Mechanics: Schrö-dinger Equation and Path Integral*, 2nd ed. (World Scientific, 2012).

[47] H.J.W. Müller–Kirsten, *Introduction to Quantum Mechanics: Schrö-dinger Equation and Path Integral*, 1st ed. (World Scientific, 2006).

[48] H.J.W. Müller–Kirsten, *Classical Mechanics and Relativity* (World Scientific, 2008).

[49] H.J.W. Müller–Kirsten, *Electrodynamics*, 2nd ed. (World Scientific, 2011).

[50] H.J.W. Müller–Kirsten, *Electrodynamics: An Introduction including Quantum Effects*, 1st ed. (World Scientific, 2004).

[51] J. Mond, *Zufall und Notwendigkeit* (Piper, 1971).

[52] N.F. Mott and H. Jones, *The Theory of the Properties of Metals and Alloys* (Oxford Univ. Press, 1958).

[53] NIST *Handbook of Mathematical Functions*, ed. by F.W.J. Olver, D.W. Lozier, R.F. Boisvart and C.W. Clerk (Cambridge Univ. Press, 2010).

[54] W. Pauli, *Z.f.Physik* **4** (1926) 81.

[55] R. Penrose, *The Road to Reality, A complete guide to the laws of the universe* (Vintage Books, 2004).

[56] E.G. Phillips, *Functions of a Complex Variable* (Oliver and Boyd, 1954).

[57] L.E. Reichl, *A Modern Course in Statistical Physics* (University of Texas Press, 1980).

[58] F. Reif, *Statistical Physics*, Berkeley Physics, Vol. 5 (1977).

[59] F. Reif, *Statistische Physik und Physik der Wärme*, transl. by A.W. Mushik, 3rd ed. (de Gruyter, 1987), of *Fundamentals of Statistical and Thermal Physics* (McGraw–Hill, 1965). Solutions to selected problems of this book have been given by: (1) R.F. Knacke, *Solutions to Problems to accompany F. Reif's Fundamentals of Statistical and Thermal Physics* (McGraw Hill, 1965), and (2) K.-P. Charlé and H.U. Zimmer, *Aufgaben zur Statistischen Physik und Theorie der Wärme mit Rechenweg und Lösungen* (W. de Gruyter, 1979).

[60] O.K. Rice, *Statistical Mechanics, Thermodynamics and Kinetics* (W.H. Freeman and Co., 1967).

[61] E.G. Richardson, *Sound*, 5th ed. (Arnold, 1953).

[62] P.C. Riedl, *Thermal Physics* (Macmillan Press, 1976).

[63] G.S. Rushbrooke, *Introduction to Statistical Mechanics* (Oxford University Press, 1949).

[64] O. Sakur, *Annalen d. Physik* **36** (1911) 958.

[65] E. Schrödinger, *Statistical Thermodynamics* (Cambridge Univ. Press, 1952).

[66] D.V. Schroeder, *An Introduction to Thermal Physics* (Oxford University Press, 2021). This book explains the fundamentals at great length and contains many problems.

[67] G. Schubert, *Z. Naturforsch.* **1** (1946) 113, **2** (1947) 250.

[68] S. Schweber, *An Introduction to Relativistic Quantum Field Theory* (Harper and Row, 1961).

[69] J. Schwinger, *Phys. Rev.* **75** (1949) 651.

[70] W. Shockley, *Electrons and Holes in Semiconductors* (Van Nostrand, 1963).

[71] A. Sommerfeld, *Z. Phys.* **47** (1928) 1.

[72] S.G. Starling and A.J. Woodall, *Physics* (Longmans, 1950).

[73] L. Susskind and J. Lindesay, *Black Holes, Information and the String Theory Revolution* (World Scientific, 2005).

[74] N.M. Temme, *Asymptotic Methods for Integrals* (World Scientific, 2015).

[75] Ya.P. Terletskii, *Statistical Physics*, Translated from the Russian by N. Fröman (North–Holland Pub. Co., 1971).

[76] H. Tetrode *Annalen d. Physik* **38** (1912) 434.

[77] M. Toda, R. Kubo and N. Saito, *Statistical Physics I* (Springer, 1983).

[78] R.C. Tolman, *The Principles of Statistical Mechanics* (Dover Publications, 1979). For a priori Probabilities see pp. 59–70.

[79] W.G. Unruh, *Phys. Rev.* **D14** (1976) 870–892.

[80] O. Vallée and M. Soares, *Airy Functions and Applications to Physics*, 2nd ed. (Imperial College Press, 2010).

[81] B.L. Van der Waerden, On the method of saddle points, *Appl. Sci. Res.* **B2** (1950) 33.

[82] J. von Neumann, *Proc. Nat. Acad.* **18** (1932) 70, 263.

[83] G.H. Wannier, *Statistical Physics* (John Wiley, 1966). This is a very readable and highly recommendable text.

[84] V.F. Weisskopf, *The Formation of Cooper Pairs and the Nature of Superconducting Currents*, CERN 79-12 (yellow report).

[85] R.E. Wilde and S. Singh, *Statistical Mechanics, Fundamentals and Modern Applications* (Wiley, 1998). The reader of our text here may find it profitable to supplement his study by reading the sections of this book on the Liouville equation, ensembles and the ergodic hypothesis, and on Bose–Einstein condensation.

[86] C.S. Zasada and R.K. Pathria, *Phys. Rev.* **A14** (1976) 1269.

Index